The Pneumatic Flow Mixing Method

The Pneumatic Flow Mixing Method

Masaki Kitazume
Tokyo Institute of Technology, Tokyo, Japan

CRC Press
Taylor & Francis Group
Boca Raton London New York

CRC Press is an imprint of the
Taylor & Francis Group, an **informa** business

A BALKEMA BOOK

CRC Press
Taylor & Francis Group
6000 Broken Sound Parkway NW, Suite 300
Boca Raton, FL 33487-2742

First issued in paperback 2020

© 2017 by Taylor & Francis Group, LLC
CRC Press is an imprint of Taylor & Francis Group, an Informa business

Typeset by MPS Limited, Chennai, India

ISBN 13: 978-0-367-57424-6 (pbk)
ISBN 13: 978-1-138-02984-2 (hbk)

Library of Congress Cataloging-in-Publication Data

**Visit the Taylor & Francis Web site at
http://www.taylorandfrancis.com**

**and the CRC Press Web site at
http://www.crcpress.com**

Table of contents

Preface   xi
List of Technical Terms and Symbols   xiv

I An overview of admixture stabilization – Evolution of
pneumatic flow mixing and the scope of the book   I
  1   Introduction   1
  2   Cement admixture stabilization techniques   2
       2.1   Basic mechanism of cement admixture stabilization   2
       2.2   Classification of cement admixture stabilization techniques   2
       2.3   *In-situ* mixing techniques   4
              2.3.1   Power blender method   4
              2.3.2   Deep mixing method   4
       2.4   *Ex-situ* mixing techniques   6
              2.4.1   Premixing method   6
              2.4.2   Lightweight treated soil method
                       (Super Geo-Material lightweight soil)   6
              2.4.3   Dewatered stabilized soil method   8
              2.4.4   Granular stabilized soil method   9
  3   Development, mechanism and applications of the pneumatic
       flow mixing method   10
       3.1   Development of the method   10
       3.2   Mechanism of the method   11
              3.2.1   Air pressure distribution in pipeline   13
              3.2.2   Characteristics of soil plugs   17
              3.2.3   Required transportation distance   17
       3.3   Soil material suitable for the method   18
       3.4   Applications of the method   18
  4   Scope of the textbook   20
  References   21

2 Factors affecting strength increase   **23**
  1   Introduction   23
  2   Mechanism of cement stabilization   24
  3   Influence of various factors on the stabilization effect   25
       3.1   Influence of the characteristics of the binder   25
              3.1.1   Chemical composition of the binder   25
              3.1.2   Type of binder   28

		3.1.3 Type of mixing water	28
		3.1.4 Type of additives	32
	3.2	Influence of the characteristics and conditions of soil	33
		3.2.1 Soil type	33
		3.2.2 Grain size distribution	35
		3.2.3 Humic acid	35
		3.2.4 Organic content	36
		3.2.5 Water content	40
	3.3	Influence of the mixing conditions	42
		3.3.1 Quantity of cement	42
		3.3.2 Mixing time	43
		3.3.3 Time and duration of mixing, and molding process	44
	3.4	Influence of the curing conditions	46
		3.4.1 Hydration	46
		3.4.2 Curing period	47
		3.4.3 Curing temperature	51
		3.4.4 Maturity	53
		3.4.5 Drying and wetting cycle	54
		3.4.6 Overburden pressure	55
		3.4.7 Soil disturbance/compaction	56
4		Prediction of strength	60
References			62
3	**Engineering properties of stabilized soils**		**67**
1		Introduction	67
2		Properties of stabilized soil mixture before hardening	68
	2.1	Physical properties	68
		2.1.1 Change in consistency of the soil-binder mixture before hardening	68
	2.2	Mechanical properties (strength characteristics)	69
		2.2.1 Change in flow value	69
		2.2.2 Change in shear strength	70
		2.2.3 Stress – strain curve	70
	2.3	Mechanical properties (consolidation characteristics)	70
3		Properties of stabilized soil after hardening	76
	3.1	Physical properties	76
		3.1.1 Change in water content	76
		3.1.2 Change in density	77
		3.1.3 Change in consistency	79
	3.2	Mechanical properties (strength characteristics)	79
		3.2.1 Stress – strain curve	79
		3.2.2 Strain at failure	81
		3.2.3 Internal friction angle and undrained shear strength	81
		3.2.4 Residual strength	84

		3.2.5	Modulus of elasticity (Young's modulus)	84
		3.2.6	Poisson's ratio	85
		3.2.7	Dynamic property	85
		3.2.8	Creep strength	87
		3.2.9	Cyclic strength	87
		3.2.10	Tensile and bending strengths	90
		3.2.11	Long-term strength	92
		3.2.12	Coefficient of horizontal stress at rest	100
	3.3		Mechanical properties (consolidation characteristics)	101
		3.3.1	Void ratio – consolidation pressure curve and consolidation yield pressure	101
		3.3.2	Coefficient of volume compressibility	103
		3.3.3	Coefficient of consolidation	104
		3.3.4	Coefficient of permeability	106
		3.3.5	Secondary compression	108
	3.4		Environmental properties	108
		3.4.1	Elution of contaminant	108
		3.4.2	Elution of hexavalent chromium (chromium VI) from stabilized soil	113
		3.4.3	Resolution of alkali from stabilized soil	116
		3.4.4	Resolution of dioxin from stabilized soil	116
4			Properties of stabilized soil subjected to disturbance/compaction	117
	4.1		Physical properties	117
		4.1.1	Change in consistency	117
	4.2		Mechanical properties (strength characteristics)	119
		4.2.1	Influence of soil disturbance	119
		4.2.2	Influence of soil disturbance and compaction	121
5			Engineering properties of cement-stabilized soil produced *in-situ*	122
	5.1		Flow value of field stabilizedsoil	122
	5.2		Mixing degree of field stabilized soil	122
	5.3		Effect of transportation distance	125
	5.4		Effect of placement	127
		5.4.1	Effect of amount of cement on strength	127
	5.5		Heterogeneity of dredged soil	129
	5.6		Property of stabilized ground	131
6			Summary	133
	6.1		Properties of stabilized soil mixture before hardening	133
		6.1.1	Physical properties	133
		6.1.2	Mechanical properties (strength characteristics)	134
		6.1.3	Mechanical properties (consolidation characteristics)	134
	6.2		Properties of stabilized soil after hardening	135
		6.2.1	Physical properties	135
		6.2.2	Mechanical properties (strength characteristics)	135
		6.2.3	Mechanical properties (consolidation characteristics)	137
		6.2.4	Environmental properties	137

	6.3	Properties of stabilized soil subjected to disturbance/ compaction		138
	6.3.1	Physical properties		138
	6.3.2	Mechanical properties (strength characteristics)		138
	6.4	Engineering properties of field cement-stabilized soil		138
	6.4.1	Flow value of field stabilized soil		138
	6.4.2	Effect of transportation distance		139
	6.4.3	Effect of placement		139
	6.4.4	Heterogeneity of dredged soil		139
	6.4.5	Property of stabilized ground		139
References				139

4   Applications of the pneumatic flow mixing method   **143**

1	Introduction		143
2	Improvement purposes and applications		143
	2.1	Applications of the method		143
3	Selected case histories of the method in Japan		145
	3.1	A field test on long-distance transport (field test)		145
	3.1.1	Outline of project		145
	3.1.2	Design and stabilization work		147
	3.2	Shallow layer construction at Nanao Port		150
	3.2.1	Outline of project		150
	3.2.2	Design and stabilization work		150
	3.3	Field test on the strength of stabilized soil placed underwater		151
	3.3.1	Outline of project		151
	3.3.2	Design and stabilization work		152
	3.4	Backfill in deep water		154
	3.4.1	Outline of project		154
	3.4.2	Design and stabilization work		154
	3.5	Land reclamation for Central Japan International Airport		155
	3.5.1	Outline of project		155
	3.5.2	Design and stabilization work		155
	3.5.3	Strength of the stabilized ground		157
	3.5.4	Long-term strength and water content of the stabilized ground		157
	3.6	Land reclamation for Tokyo/Haneda International Airport		160
	3.6.1	Outline of project		160
	3.6.2	Design and stabilization work		161
	3.7	Land reclamation using converter slag		164
	3.7.1	Outline of project		164
	3.7.2	Stabilization work		164
	3.8	Backfill behind breakwater – for settlement reduction (field test)		166
	3.8.1	Outline of project		166
	3.8.2	Design and stabilization work		166

3.9   Backfill behind breakwater – for settlement reduction
        (field test)   167
        3.9.1   Outline of project   167
        3.9.2   Design and stabilization work   169
References   171

**5   Equipment, construction, and quality control
and assurance   173**
1   Introduction   173
2   Equipment   173
      2.1   System and specifications   173
      2.2   Air pressure feed system   176
      2.3   Binder supplier system   177
      2.4   Pipeline   180
      2.5   Placement equipment   184
      2.6   Control equipment   184
3   Construction procedure   186
      3.1   Preparation of site   186
      3.2   Field trial test   186
      3.3   Construction work   186
              3.3.1   Remolding and water content control   186
              3.3.2   Injection of binder and transportation of soil   186
              3.3.3   Placement of stabilized soil   188
4   Quality control   191
      4.1   Quality control before production   191
              4.1.1   Soil property   191
              4.1.2   Pipeline length   191
      4.2   Quality control during execution   191
              4.2.1   Material control   191
              4.2.2   Transportation control   191
              4.2.3   Placement control   192
      4.3   Quality assurance   197
              4.3.1   Shape of stabilized soil ground   197
              4.3.2   Strength of stabilized soil ground   197
              4.3.3   Environmental impact during placement   197
              4.3.4   Water quality control   201
              4.3.5   Elution of hexavalent chromium (chromium VI)
                       from stabilized soil   201
References   203

6   Geotechnical design of stabilized soil ground   205
1   Introduction   205
2   Design strength   205
      2.1   Relationships of laboratory strength, field strength
              and design strength   205
      2.2   Design flow for field and laboratory stabilized soil
              strengths and mixing condition   207

3   Geotechnical design   208
        3.1   Earth pressure of stabilized soil ground with infinite width   208
                3.1.1   Earth pressure before hardening   208
                3.1.2   Earth pressure after hardening   209
        3.2   Earth pressure of stabilized soil ground with a finite width   211
        3.3   Bearing capacity of stabilized soil ground   213
        3.4   Liquefaction of stabilized soil   213
        3.5   Soil volume design   213
References   215

A   Japanese laboratory mix test procedure   217
1   Introduction   217
2   Testing equipment   217
        2.1   Equipment for making specimen   217
                2.1.1   Mold   217
                2.1.2   Mixer   218
                2.1.3   Binder mixing tool   218
        2.2   Soil and binder   219
                2.2.1   Soil   219
                2.2.2   Binder   220
3   Making and curing of specimens   220
        3.1   Mixing materials   220
        3.2   Making a specimen   221
        3.3   Curing   222
        3.4   Specimen removal   223
4   Report   223
5   Use of specimens   225
References   227
Subject index   229

Preface

It is an obvious truism that, structures should be constructed on a stiff ground to ensure their stability and negligible deformation. Along with the development of society and concentration of population to urban areas, the ground conditions of construction sites, however, have become worse than ever during recent decades throughout the world. This situation is especially pronounced in Japan, where many construction projects are conducted on soft alluvial clay ground, land reclaimed ground with dredged soils, highly organic soil ground, loose sandy ground and so on. It is often encountered to such a soft soil when any type of infrastructures is constructed, in which large ground settlement and stability failure are concerned. Apart from these clayey or highly organic soils, loose sand deposits under the water table can cause serious problems of liquefaction under seismic conditions. A lot of *in-situ* ground improvement techniques have been developed to improve physical and mechanical properties of the soft soil in order to cope with these problems.

A huge amount of soft soil is dredged at many ports annually to maintain sea routes and berths. A huge amount of construction waste soil and industrial by-product are also produced at construction sites and industry plants. These soils used to be dumped at disposal areas constructed in coastal area and mountainous area. It is becoming difficult to construct any disposal areas for these soils, because of environmental restrictions and economic reasons. It has become necessary to beneficially use these soils as construction and land reclamation materials. Several *ex-situ* soil improvement techniques have been developed for the purpose.

The cement stabilization technique is one of the common soil improvement techniques to improve the physical and mechanical properties of the original soil, in which soil and cement are mixed *in-situ* or *ex-situ* by mixing paddles or mixing blades. The shape of mixing paddles and blades as well as mixing procedure are the essential issue to ensure the uniform and desired characteristics of stabilized soil. A lot of research efforts have been paid in Japan to develop appropriate mixing machine and mixing procedure for many *in-situ* and *ex-situ* mixing techniques. As soil and cement are mixed batch by batch with small soil volume in the *ex-situ* mechanical mixing technique using mixing paddles and blades, the mechanical mixing techniques are not effective for large scale cement stabilization project. A new type of cement stabilization technique was desired for promoting beneficial use of large amount of dredged soil and surplus soils efficiently.

The pneumatic flow mixing method was developed to stabilize dredged soil and surplus soil for promoting their beneficial use in 1999. The pneumatic flow mixing method is a new type of the *ex-situ* cement stabilization techniques, in which dredged soil or surplus soil is mixed with a relatively small amount of cement without any mixing paddles and blades in a pipeline. Transporting soil in a pipeline without any air requires high pressure due to the friction generated on the inner surface of pipeline. When a relatively large amount of compressed air is injected into the pipeline, however, soil can be separated into small blocks. The separated soil block, called a 'plug', is forwarded to an outlet with the help of compressed air. The forming plug and air block functions to reduce the friction on the inner surface of pipeline and in turn can reduce the required air pressure considerably for transporting. When binder is injected into the pipeline, the soil plug and binder are thoroughly mixed by means of turbulent flow generated in the plug during transporting. The soil and binder mixture, transported and placed at the reclamation site, gains high strength rapidly so that no additional soil improvement is necessary for assuring the successful performance of superstructures on the ground in many cases. As the mixing and transporting soil and binder can be conducted in a pipeline continuously, the methods is suitable and efficient for large scale project.

The development of the pneumatic flow mixing method has been initiated by Ministry of Transport, The Fifth District Port Construction Bureau (currently Chubu Regional Development Bureau, Ministry of Land, Infrastructure, Transport and Tourism) since 1998. Many laboratory tests were carried out to investigate the physical and mechanical properties of stabilized soil, and many full scale land reclamation and backfill tests were also carried out at Nagoya Port to investigate the physical and mechanical properties of field-stabilized soil, to investigate the environmental impact to surrounding area due to the placement of stabilized soil and to develop a construction procedure, quality control and assurance. The method was applied in several land reclamation and embankment construction projects at Fushikitoyama Port, Muroran Port and Kushiro Port in 1988 to 2001. They have confirmed the high applicability of the method for an economical and rapid construction of land reclamation and embankment. The method was applied to construct large scale man-made islands for Central Japan International Airport at Nagoya in 2001 and for Tokyo Haneda International Airport in 2009.

The current book is intended to provide the state of the art and practice for pneumatic flow mixing, rather than a user-friendly manual. The book covers the factors affecting the strength increase, the engineering characteristics of stabilized soil, a variety of applications, current design procedures, execution systems and procedures, and QC/QA methods and procedures based on the experience and research efforts accumulated in Japan.

The strength of the stabilized soil is influenced by many factors, including the original soil properties; the type and amount of binder; the mixing and placement process; and curing conditions. Therefore, the process design, production with careful quality control and quality assurance are the key to the pneumatic flow mixing project. Quality assurance starts with the soil characterization of the original soil and includes various activities prior to, during, and after the production. QC/QA methods and procedures and acceptance criteria should be determined before the actual production, and their meanings should be understood precisely by all the parties involved in pneumatic flow mixing project. Contractors and practicing engineers are advised develop their own mixing machine and procedure for their site condition and design requirement. The author wishes the book to be useful for practicing engineers to understand the current state of the art and to develop their techniques and also useful for academia to find out the issues to be studied in the future.

April 2016

List of Technical Terms and Symbols

Definition of technical terms

additive	chemical material to be added to stabilizing agent for improving characteristics of stabilized soil
binder	chemically reactive material that can be used for mixing with soils to improve engineering characteristics of soils such as lime, cement, lime-based and cement-based special binders. Also referred to as stabilizer or stabilizing agent.
binder	content ratio of weight of dry binder to the dry weight of soil to be stabilized. (%)
binder factor	ratio of weight of dry binder to the volume of soil to be stabilized. (kg/m^3)
binder slurry	slurry-like mixture of binder and water
ex-situ mixing	mixing technique in which the soil is once excavated and mixed with binder in a plant or during transportation
field strength	strength of stabilized soil produced in-situ
improved ground	a region with stabilized soil and surrounding original soil
in-situ mixing	a mixing technique in which natural soil is stabilized with binder *in-situ*
laboratory strength	strength of stabilized soil produced in a laboratory
plug	soil block created in the pipeline by injecting compressed air
stabilized soil	soil stabilized by mixing with binder

List of Symbols

aw	cement content (%)
C	cement factor (kg/m^3)
c	cohesion of stabilized soil (kN/m^2)
C_c	compression index
C_s	sweling index
c_u	undrained shear strength (kN/m^2)
CV	coefficient of variation
c_v	coefficient of consolidation (m^2/day)
c_{vs}	coefficient of consolidation of stabilized soil (m^2/day)
c_{vu}	coefficient of consolidation of unstabilized soil (m^2/day)

C_α secondary compression coefficient
D pipeline diameter (m)
e void ratio
E_{50} modulus of elasticity (kN/m^2)
Fs safety factor
G shear modulus (kN/m^2)
G_0 initial shear modulus (kN/m^2)
G_c specific gravity of binder
G_{eq} equivalent shear modulus (kN/m^2)
G_s specific gravity of soil particle
G_w specific gravity of water
h_{eq} equivalent dumping ratio (%)
I_p plasticity index
k coefficient of permeability (cm/sec)
K_0 coefficient of earth pressure at rest
k_{30} ground reaction coefficient (kN/m^2)
k_h horizontal seismic coefficient
k'_h apparent horizontal seismic coefficient
L_i ignition loss (%)
M mixing ratio
M maturity
m_v coefficient of volume compressibility (m^2/kN)
m_{vs} coefficient of volume compressibility of stabilized soil (m^2/kN)
m_{vu} coefficient of volume compressibility of unstabilized soil (m^2/kN)
N_c bearing capacity factor of soil
N_γ bearing capacity factor of soil
N_q bearing capacity factor of soil
p'_0 effective earth pressure coefficient at rest (kN/m^2)
p_a active earth pressure (kN/m^2)
pH   potential hydrogen
p_t earth pressure of liquid state stabilized soil (kN/m^2)
p_y consolidation yield pressure (pseudo pre-consolidation pressure) (kN/m^2)
q_c cone penetration resistance (kN/m^2)
q_d ultimate bearing capacity (kN/m^2)
q_f bearing capacity (kN/m^2)
Q_A flow volume of air (g/m^3)
Q_L flow volume of soil (g/m^3)
q_u unconfined compressive strength (kN/m^2)
q_{uck} design standard strength (kN/m^2)
q_{uf} unconfined compressive strength of field stabilized soil (kN/m^2)
$\overline{q_{uf}}$ average unconfined compressive strength of field stabilized soil (kN/m^2)
q_{ul} unconfined compressive strength of laboratory stabilized soil (kN/m^2)

$\overline{q}_{ul}$	average unconfined compressive strength of laboratory stabilized soil (kN/m^2)
q_y	yield bearing capacity (kN/m^2)
s_u	shear strength (kN/m^2)
S	settlement (m)
SS	suspended solids (mg/L)
t	curing period (day)
t_c	curing period (day)
T_c	curing temperature (°C)
u	plug speed (m/sec)
V_c	volume of cement slurry (m^3)
V_{cw}	volume of water in binder slurry (m^3)
V_s	volume of soil (m^3)
V_{soil}	volume of soil (m^3)
V_v	volume of void and water (m^3)
V_{w2}	volume of water added on transporting barge (m^3)
w	water content (%)
w	surcharge (kN/m^2)
W/C	water to cement ratio of cement slurry (%)
W/C	water to cement ratio of stabilized soil defined as the total water containing in both soil and cement slurry against the dry weight of binder (%)
w/w_L	water content ratio
W_c	cement factor (kg/m^3)
W_c	dry weight of cement (kg)
w_i	initial water content (%)
w_L	liquid limit (%)
w_n	natural water content (%)
w_P	plastic limit (%)
w_t	total water content (%)
α	cement factor (kg/m^3)
α	angle of slip surface (°)
β	angle of back fill (°)
δ	friction angle at wall surface (°)
ε_f	axial strain at failure (%)
ϕ'	internal friction angle (°)
γ	shear strain (%)
γ	unit weight (kN/m^3)
γ'	effective unit weight (kN/m^3)
γ_w	unit weight of water (kN/m^3)
λ	strength ratio of q_{uf}/q_{ul}
λ	water to soil ratio required for cement hydration (%)
μ	Poisson's ratio
ν	viscocity of plug (P)
θ	resultant angle of seismic coefficient (°)
θ	curing temperature (°)
ρ_c	density of cement (g/cm^3)
ρ_A	density of air (g/cm^3)

ρ_L	density of soil (g/cm^3)
ρ_s	density of stabilized soil (g/cm^3)
ρ_{soil}	density of original soil (g/cm^3)
ρ_w	density of water (g/cm^3)
σ	standard deviation
σ	confining pressure (kN/m^2)
σ_{tb}	tensile strength measured by bending test (kN/m^2)
σ_{td}	tensile strength measured by simple tension test (kN/m^2)
σ_{ts}	tensile strength measured by split test (kN/m^2)
ψ	angle of wall to the vertical (°)
ζ	angle of failure surface to horizontal (°)

An overview of admixture stabilization – Evolution of pneumatic flow mixing and the scope of the book

1 INTRODUCTION

Many man-made islands have been constructed in Japan to obtain enough new land area for airports, electrical power plants, manufacturing plants, residential areas and so on. These islands require a huge amount of soil with appropriate soil properties for land reclamation, which used to be excavated from mountains, rivers and seabed. Recently it is becoming difficult to obtain such soils at a reasonable cost, because of environmental protection restrictions and economic reasons, which requires the use of inappropriate soil as a reclamation material. A huge amount of soft soil is dredged at many ports annually to maintain sea routes and berths. A huge amount of surplus soil and by-produce are also produced at construction sites and industry plants. These soils used to be dumped at disposal areas constructed in coastal areas, because disposal offshore is strictly prohibited in Japan. It is becoming difficult to construct any disposal areas for dredged soil and surplus soil, because of environmental protection restrictions and economic reasons. These circumstances have promoted the beneficial use of dredged soft soil and surplus soil as a land reclamation material, which in turn can prolong the service life of disposal areas. As the dredged soil and surplus soil usually have a high water content, of the order of a hundred per cent, reclaimed lands constructed with the soil are so weak and highly compressible that ground improvement is necessary for constructing structures on the reclaimed land to assure their stability and to reduce their settlement. The vertical drain method is one of the most frequently used ground improvement techniques to improve such soft soil deposits. However, the method requires a relatively long period to complete the consolidation of the ground.

Many studies and much research have been carried out to investigate appropriate soil improvement methods and to establish a recycling system for the soft soil. The pneumatic flow mixing method, one of the admixture stabilization techniques, was developed to stabilize dredged soil and surplus soil for promoting their beneficial use in Japan in 1999, in which soil and a chemical binder are mixed in the pipeline by the help of the turbulent flow generated in the soil plug during transportation by compressed air. As this method has many benefits – rapid and large scale execution can be conducted with low cost – it has been applied to many land reclamation projects, backfilling behind earth retaining wall projects and shallow stabilization projects using dredged soils and surplus soils.

In the following sections, the classification, development, basic mechanism and application of the pneumatic flow mixing method will be explained briefly.

2 CEMENT ADMIXTURE STABILIZATION TECHNIQUES

2.1 Basic mechanism of cement admixture stabilization

The basic mechanism of cement stabilization is illustrated in Figure 1.1, which consists of four steps: the reduction of water content, improvement of physical properties, cement hydration hardening, and pozzolanic reaction hardening. The water content of the original soil is decreased by the hydration of a binder and subsequent water absorption process. The ion exchange reaction modifies the physical property of the original soil and results in a decrease of plasticity of the soil. This effect is utilized in the improvement of base and sub-base soils by mixing with small amounts of lime or cement for road construction, where the change of consistency of the base and sub-base soils makes compaction easier and more effective. The formation of cement hydration products and pozzolanic reaction products provide an increased strength to the soil and cement mixture. The pneumatic flow mixing method is mostly based on the latter two reactions to increase the strength of the soil.

2.2 Classification of cement admixture stabilization techniques

The success of the deep mixing method (Kitazume & Terashi, 2013) for soil stabilization has encouraged the construction industry to develop various types of cement

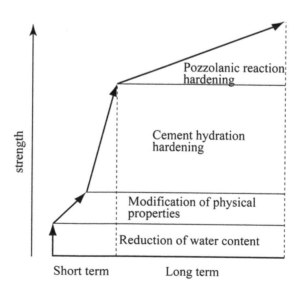

Figure 1.1 Mechanism of cement stabilization.

admixture stabilization techniques in Japan. The currently available cement admixture stabilization techniques can be classified into *in-situ* mixing and *ex-situ* mixing, as shown in Table 1.1 (after Coastal Development Institute of Technology, 2008b). The *in-situ* mixing techniques are developed to improve the physical and mechanical properties of the original soil for assuring the successful performance of superstructures on the ground. In the *in-situ* mixing techniques, natural soil is stabilized with binder *in-situ* by means of mechanical mixing and/or high pressure injection mixing. The *in-situ* mixing techniques can be subdivided into surface and shallow stabilization; mid-depth stabilization; and deep stabilization, depending upon the depth and purpose of improvement. The *ex-situ* mixing techniques have been developed to enhance the beneficial use of dredged soils, inappropriate soils and construction surplus soils. These techniques are intended to provide additional characteristics – such as better liquefaction resistance, smaller density, smaller volume compressibility or high strength – to the original soil. In the *ex-situ* mixing techniques, the soils are, once excavated, mixed with binder in a plant or during transportation to a disposal site or reclamation site. The *ex-situ* mixing techniques can be further subdivided into mixing during transportation and batch plant mixing, depending upon where soil and binder are mixed. Several *in-situ* and *ex-situ* admixture stabilization techniques will be briefly introduced in the next section.

Table 1.1 Classification of admixture stabilization techniques (after Coastal Development Institute of Technology, 2008b).

Place of mixing		Type of mixing	Method	Application
In-situ	Surface and shallow stabilization	mechanical mixing	surface treatment, shallow stabilization	working platform on soft ground
	Mid depth stabilization	mechanical mixing	mid-depth mixing	stability, settlement reduction, excavation support, seepage shutoff, etc
	Deep stabilization	mechanical mixing high pressure injection hybrid of above two	deep mixing	
Ex-situ	Mixing during transportation	mixing on belt conveyor	pre-mixing	improve liquefaction resistance of soil
		mixing in pipeline	pipe mixing	reduce compressibility of high water content soil
	Batch plant mixing	mechanical mixing	pre-mixing	improve liquefaction resistance of soil
		mechanical mixing	lightweight geo-material	density control of fill material
		mechanical mixing and high pressure dewatering	dewatered stabilized soil	alternative for sand and gravel

Figure 1.2 Power blender method (Power Blender Method Association, 2006).

2.3 *In-situ* mixing techniques

2.3.1 *Power blender method*

The power blender method is one of the mid-depth *in-situ* mixing techniques. The mixing machine of the method installs a large number of small stirring wings attached to the trencher, as shown in Figure 1.2 (Power Blender Method Association, 2006). In the method, the natural ground is cut and mixed with binder by rotating the stirring wings *in-situ*, where powder or slurry form binder is injected from the stirrer tip. The method can stabilize ground to a depth of about 13 m for the purpose of improving stability, reducing ground settlement and preventing liquefaction. As the trencher is installed on the extension arm of backhoe, the base machine can be placed on a firm layer far from the soft ground to be stabilized.

2.3.2 *Deep mixing method*

The deep mixing method is a deep *in-situ* soil stabilization technique using cement and/or lime as a binder, which forms a stiff, artificially cement-stabilized soil of various shapes, e.g. columns, walls, panels, grids or blocks, in order to improve foundation ground (Kitazume & Terashi, 2013). A large machine with several rotating shafts and mixing blades is used to supply powder or slurry form binder into the ground and mix soil and binder *in-situ*, as shown in Figure 1.3. The applications of the method have various purposes, such as reducing ground settlement, increasing bearing capacity of the ground, increasing stability, preventing ground liquefaction, reducing active earth pressure, cutting off ground water, and increasing piles' lateral resistance.

(a) For work on land.

(b) For marine work.

Figure 1.3 Mixing machine of deep mixing method.

2.4 *Ex-situ* mixing techniques

2.4.1 *Premixing method*

The premixing method is one of the *ex-situ* admixture stabilization techniques where a small amount of binder and chemical additives are mixed with sandy material to obtain liquefaction-free material for land reclamation (Zen et al., 1987, Coastal Development Institute of Technology, 2003, 2008c). The basic principle of the method is to prevent liquefaction by a cementation effect between the sand particles and the binder. In the case where soil has a certain degree of cohesion by the cementation effect, the shear strength does not decrease to zero and liquefaction does not take place even when pore water pressure is generated up to the overburden pressure. Sand, binder and chemical additive (e.g. a separation inhibitor) are mixed either in a batching plant or on a belt conveyor. The stabilized soil is transported and placed at the designated area to construct reclaimed ground (Figure 1.4). This method was applied to the restoration works of quays at Kobe Port where a concrete-type quay wall was damaged by the Great Hanshin earthquake in 1995 (Kitazume, 2010).

2.4.2 *Lightweight treated soil method (Super Geo-Material lightweight soil)*

The lightweight treated soil method (Super Geo-Material lightweight soil) is one of the *ex-situ* admixture stabilization techniques which was developed for the purpose of reducing residual and uneven ground settlement, decreasing active earth pressure, preventing lateral displacement and improving seismic resistance. In the method, dredged soil is mixed with binder and also either air foam or expanded polystyrene (EPS) beads of 1 to 3 mm in diameter in order to manufacture high quality soil, having high strength and a low unit weight of 11 and $15\,kN/m^3$ (Tsuchida & Egashira, 2004, Coastal Development Institute of Technology, 2008a). Helped by its high strength and

Figure 1.4 Premixing method (by the courtesy of Dr Yamazaki).

light weight characteristics, the stabilized soil is used for landfill, or backfill behind a quay wall or earth retaining structure, where the overburden pressure and active earth pressure can be reduced considerably. The method has been applied to backfill behind a quay wall, reinforcement of an existing structure, and an embankment on soft ground (Figure 1.5). This method was successfully applied to the restoration works of quays at Kobe Port damaged by the Great Hanshin earthquake in 1995 (Kitazume, 2010) and the landfill of a sea wall at the D runway of Tokyo Haneda International Airport (Mizukami & Matsunaga, 2015).

(a) Special barges for marine works.

(b) Placement of stabilized soil for land reclamation.

Figure 1.5 Lightweight treated soil method.

(a) High pressure compressor.

(b) Crashed dewatered stabilized soil.

Figure 1.6 Dewatered stabilized soil method.

2.4.3 Dewatered stabilized soil method

In order to produce a high strength and compacted stabilized soil with low water content, several dewatered stabilized soil methods were developed in which soil is mixed with binder and dewatered at high compressive pressure of the order of 1 to 4 MN/m^2 as shown in Figure 1.6(a). By the procedure, the stabilized soil having high strength of the cone penetration resistance, q_c, of about 400 to 600 kN/m^2 can be manufactured. The stabilized and compressed soil is usually crushed to granular material as shown in Figure 1.6(b), which can be used as subgrade and roadbed materials and as fill material.

(a) Batch plant.

(b) Granular stabilized soil.

Figure 1.7 Granular stabilized soil method (Mud Recycling System) (Source: http://www.penta-ocean. co.jp/english/business/envi/dirt_recycle.html)

2.4.4 *Granular stabilized soil method*

The granular stabilized soil method manufactures grains of a few millimetres to about 10 mm in diameter by mixing soft soil with binder and polymer or additives in a granulating mixer. Organic water-soluble polymers and inorganic agents are usually used for the polymer (polyelectrolyte), which functions to facilitate the granulation process by the aggregation effect. Iron and steel slags, coal ash or paper sludge ash is used as an additive for soils having a high water content to facilitate the granulation process by reducing the mixture's water content. The granular stabilized soil, having a high strength, is manufactured after several days' to several weeks' curing. It can be used as subgrade and roadbed materials and as fill material. Figure 1.7(a) shows one example of the batching plant for the method (Mud Recycling System), where dredged soil with a high water content is mixed with fly ash, water-soluble polymer and cement. Figure 1.7(b) shows the stabilized soil grains, which have a diameter of about 1 mm, a unit weight of 14 kN/m^3; the internal friction angle is about 40 degrees.

Figure 1.8 Barges for the pneumatic flow mixing method.

3 DEVELOPMENT, MECHANISM AND APPLICATIONS OF THE PNEUMATIC FLOW MIXING METHOD

3.1 Development of the method

The pneumatic flow mixing method, Figure 1.8, is one of the *ex-situ* stabilization techniques, in which dredged soil or surplus soil is mixed with a relatively small amount of cement in a pipeline. The development of the pneumatic flow mixing method has been initiated by the Ministry of Transport, The Fifth District Port Construction Bureau (currently the Chubu Regional Development Bureau, Ministry of Land, Infrastructure, Transport and Tourism) since 1998, where many laboratory tests were carried out to investigate the mechanical properties of stabilized soil, and many full-scale land reclamation and backfill tests were also carried out at Nagoya Port to investigate the mechanical properties of field-stabilized soil and to develop construction procedures and quality control and assurance. The Bureau summarized the laboratory and field test results and published a technical manual in 1999 (Ministry of Transport, The Fifth District Port Construction Bureau, 1999). This has encouraged many contractors to develop their own techniques of the method, as later shown in Chapter 5. The technical development study and investigation committee on the pneumatic flow mixing method was established in 1998 to summarize the accumulated test results, knowledge and know-how for promoting the method. The committee published the technical manual in 2001 (Coastal Development Institute of Technology, 2001), which was revised in 2008 (Coastal Development Institute of Technology, 2008b) in accordance with the revised Port and Harbour Design Standard (Ministry of Land, Infrastructure, Transport and Tourism, 2007). The method was applied in several land reclamation and embankment

Figure 1.9 Pneumatic flow mixing method at the Central Japan International Airport construction project.

construction projects at Fushikitoyama Port, Muroran Port and Kushiro Port in 1988 to 2001. In Fushikitoyama Port, a total of 76,750 m³ dredged soil was stabilized with 60 or 80 kg/m³ cement for constructing an embankment. In Kushiro Port, a total of 205,200 m³ of dredged soil was stabilized with 70 to 80 kg/m³ cement for land reclamation. These projects have revealed the high applicability of the method as an economical and rapid construction technique for embankment and land reclamation. The method was applied to construct a huge scale man-made island for Central Japan International Airport at Nagoya in 2001 (Kitazume & Satoh, 2003, Kitazume, 2004, Kitazume & Satoh, 2005), where about 8.2 million m³ of dredged soil was stabilized for land reclamation, see Figure 1.9 (see Chapter 4). After the construction of the airport, the method has been applied to many marine construction and on land construction projects, in which their purpose was backfilling behind sea revetment (Kobayashi et al., 2001, Yamauchi et al., 2011, Yamagoshi et al., 2013) and constructing shallow stabilized soil layer (Watanabe, 2005).

In 2008, the expansion project of Tokyo Haneda International Airport commenced, where a total of about 4.7 million m³ of dredged soil was stabilized to construct a part of the man-made island for the fourth runway, D runway, as shown in Figure 1.10 (Mizukami & Matsunaga, 2015).

3.2 Mechanism of the method

Transporting soil in a pipeline without any air requires high pressure due to the friction generated on the inner surface of pipeline. When a relatively large amount of

Figure 1.10 Pneumatic flow mixing method at the Tokyo Haneda International Airport construction project.

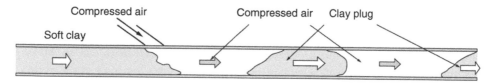

Figure 1.11 Schematic view of plug flow (Ministry of Transport, The Fifth District Port Construction Bureau, 1999).

compressed air is injected into the pipeline, however, soil can be separated into small blocks, as schematically shown in Figure 1.11 (Ministry of Transport, The Fifth District Port Construction Bureau, 1999). The separated soil block, called as 'plug', is forwarded to an outlet with the help of compressed air. The forming plug and air block functions to reduce the friction and in turn can reduce the required air pressure considerably for transporting. The formation of the plug is dependent upon the ratio of soil and air in the mix, and the pile diameter, as shown in Figure 1.12 and Equation 1.1 (Akagawa, 1980).

When binder is injected into the pipeline, the soil plug and binder are thoroughly mixed by means of turbulent flow generated in the plug during transporting. The soil and binder mixture, transported and placed at the designated reclamation site, gains high strength rapidly so that no additional soil improvement is necessary for assuring the successful performance of superstructures on the stabilized soil ground in

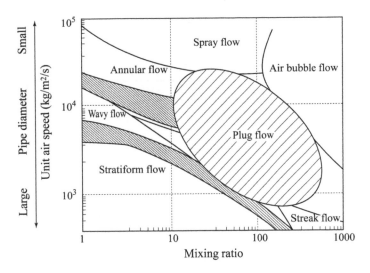

Figure 1.12 Generation condition of plug flow (Akagawa, 1980).

many cases (Coastal Development Institute of Technology, 2008b, Kitazume & Satoh, 2003).

$$M = \frac{\rho_L \times Q_L}{\rho_A \times Q_A} \tag{1.1}$$

where
M: mixing ratio
Q_A: flow volume of air (g/m^3)
Q_L: flow volume of soil (g/m^3)
ρ_A: density of air (g/cm^3)
ρ_L: density of soil (g/cm^3)

3.2.1 *Air pressure distribution in pipeline*

The required injected air pressure is dependent upon many factors such as the properties of soil, injected air volume, pipeline diameter and pipeline length. The maximum air pressure is also determined by the air resistant capacity of the facility and pipeline. In many cases, the maximum air pressure is determined to be around 600 kN/m^2. An example of air pressure distribution measured in the field test is shown in Figure 1.13 along the transportation distance measured from the inlet (Ministry of Transport, The Fifth District Port Construction Bureau, 1999). In the field test, soft marine clay with a water content of about 100% was transported by the method with three different transported soil volume rates. The figure shows that a relatively large decrease in the air pressure occurred between at the pneumatic barge, P_0, and at the stabilizing agent supplier barge, P_1, irrespective of the test condition, which was probably because the

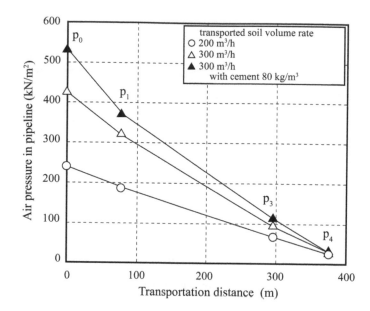

Figure 1.13 Air pressure distribution in the pipeline (Ministry of Transport, The Fifth District Port Construction Bureau, 1999).

clay in the pipeline was still in too transitional a condition to form the plug. After passing at the point P_1, the air pressure decreased almost linearly to almost zero at the outlet, P_4, with the transportation distance. The decrease in the air pressure was thought to be attributable to the inner surface friction on the pipeline. The effect of soil volume rate on the air pressure distribution is also shown in the figure, in which the higher air pressure is required at the inlet, P_0, with increasing soil volume rate. In the test conditions, the required air pressure should be increased about 200 kN/m^2 when the soil volume rate increased from 200 m^3/h to 300 m^3/h. The figure shows that the addition of cement causes the pressure increase of about 100 kN/m^2 at the inlet, P_0, because the cohesion and adhesion of the soil and cement mixture becomes larger than the soil alone. In current practice, the inlet air pressure of 400 to 500 kN/m^2 is frequently adopted after considerations of these test results, as well as the pressure capacity of facility and pipeline.

The air pressure at the inlet is influenced by many factors, which are introduced briefly below. Figure 1.14(a) shows the effects of the flow value of stabilized soil on the air pressure at the inlet (Kitazume et al., 2007). There is no clear relationship between the flow value and the air pressure, but the air pressure ranges from about 130 to 450 kN/m^2. According to Figure 1.14(a), the gradient of air pressure decreases with the flow value of stabilized soil, which indicates that the air pressure at an inlet may decrease with the flow value. Figures 1.14(b) and 1.14(c) show the effect of the transportation distance. There is a lot of scatter in the relationship between the air pressure and the transportation distance. However, the air pressure slightly increases with the transportation distance. Figure 1.14(d) shows the effect of the cement factor.

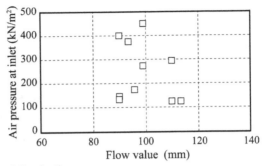

(a) Flow value of stabilizeds oil.

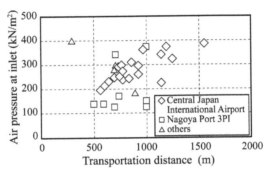

(b) Pipeline length.

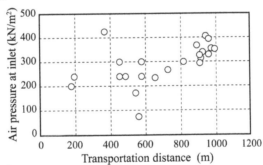

(c) Pipeline length.

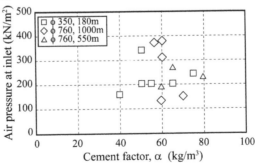

(d) Cement factor.

Figure 1.14 **Effects of characteristics of stabilized soil and transportation distance on air pressure at the inlet (Kitazume et al., 2007).**

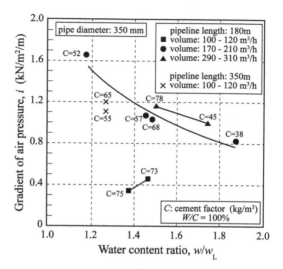

(a) Effect of water content ratio, w/w_L.

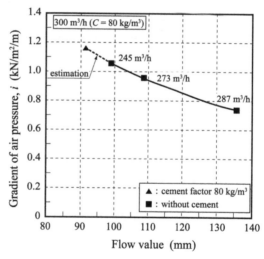

(b) Effect of flow value.

Figure 1.15 **Effect on gradient of air pressure in the pipeline (Ministry of Transport, The Fifth District Port Construction Bureau, 1999).**

According to Figure 1.14, the air pressure ranges from about 100 to 400 kN/m² irrespective of the flow value, the transportation distance and the cement factor in many cases.

Figure 1.15 shows the effects of the water content ratio, w/w_L and the flow value of soil cement mixture on the gradient of the air pressure in the pipeline, i (Ministry of Transport, The Fifth District Port Construction Bureau, 1999). The gradient, i,

Table 1.2 Characteristics of soil plugs measured in field tests (Ministry of Transport, The Fifth District Port Construction Bureau, 1999).

	Case 1	Case 2	Case 3
Test condition			
soil volume (m³/h)	210	296	170
water content (%)	132.5	117.2	96.2
liquid limit (%)	70.5	77.6	81.5
cement factor (kg/m³)	38	78	52
Test results	mean (range)	mean (range)	mean (range)
plug speed (m/s)	10.9 (1.6–25.0)	11.9 (1.5–25.0)	12.8 (1.9–25.0)
plug volume (m³)	0.41 (0.23–0.52)	0.36 (0.25–0.45)	0.30 (0.18–0.36)
plug length (m)	4.3 (0.23–5.4)	3.7 (2.6–4.7)	3.1 (1.9–3.8)
plug interval (s)	7.1 (1.3–30.3)	4.4 (0.5–18.2)	6.4 (0.6–29.0)

decreases with the water content ratio and the flow value, which indicates the required air pressure at the inlet may decrease.

3.2.2 Characteristics of soil plugs

An example of the characteristics of soil plugs measured in field tests are summarized in Table 1.2 (Ministry of Transport, The Fifth District Port Construction Bureau, 1999). A soil plug with an average volume of 0.36 m³ is transported every about 6 seconds at an average speed of about 12 m/s. As the soil plug is transported at very high speed in the pipeline, a turbulent flow is generated within the soil plug due to the friction on the inner surface of the pipeline. The turbulent flow functions to mix the soil and binder thoroughly. Thorough mixing can be obtained in the Reynolds number, $Re = uD/v$ of 500 to 3,000, where D is the pipeline diameter and u and v are the speed and viscosity of soil plug respectively.

3.2.3 Required transportation distance

It is obvious that a certain run of transportation distance is required to ensure throughout mixing, since the soil and binder are mixed in the pipeline during transportation. The required transportation distance may be dependent on many factors, such as the diameter of pipeline, the pressure and volume of air pressure, the volume and characteristics of soil, and the volume of binder. However, there are little background data regarding the minimum transportation distance to ensure throughout mixing, which will be explained in Chapter 5. The minimum transportation distance to ensure sufficient mixing in the pipeline might be determined empirically in each project. It is known that at least 50 to 100 m of transportation distance is necessary for sufficient mixing, while accumulated case histories suggest a transportation distance exceeding about 200 m is necessary in many cases, as shown later in Chapter 4, in Figure 4.6.

Table 1.3 Soil properties suitable for the pneumatic flow mixing method (Coastal Development Institute of Technology, 2008b).

Soil type		Water content	Applicability	Evaluation Binder content	Pumping ratio
cohesive soil	sand content 30–50%	high	good	middle	somewhat low
		low	good	middle	low
	sand content <30%	higher than 200% (higher than 2.8 w_L)	not good	–	–
		110 to 200% (1.5 to 2.8 w_L)	good	rich	fair
		90 to 110% (1.3 to 1.5 w_L)	good	middle	fair
		70 to 90% (1.0 to 1.3 w_L)	good	middle	somewhat low
		50 to 70% (0.7 to 1.0 w_L)	good	middle	low
		lower than 50% (lower than 0.7 w_L)	not good	–	–

3.3   Soil material suitable for the method

The pneumatic flow mixing method can be applicable not only to dredged soft soil but also to surplus soil and subsoil produced from construction sites. Soil properties suitable for the method are summarized in Table 1.3 (Coastal Development Institute of Technology, 2008b), in which cohesive soil with a sand particle content of less than 30% and a water content of 90 to 110% (approximately 1.3–1.5 times the liquid limit) is the most suitable for the method. On the other hand, cohesive soil with a sand particle content of 30 to 50% and a comparatively low water content, or cohesive soil with a sand particle content of less than 30% and a water content of 50 to 70% (approximately 0.7–1.0 times the liquid limit) are not suitable for the method, but can be available if they are remolded throughout by adding additional water to reduce their cohesion.

3.4   Applications of the method

The pneumatic flow mixing method has many advantages, such as making beneficial use of dredged soil and subsoil possible, obtaining any target stabilized soil strength within a short period by controlling the type and amount of binder, and conducting a rapid and large scale operation. Because of these advantages, the method has been applied to many construction projects for many improvement purposes, including land reclamation; backfilling behind sea revetments and earth retaining structures; and shallow stabilization and backfill underwater, as shown in Figure 1.16 (Ministry of Transport, The Fifth District Port Construction Bureau, 1999). For these applications, ordinary Portland cement and blast furnace slag cement type B are often used as a binder, where the cement content and target unconfined compressive strength of the stabilized soil are of the order of 50 to 70 kg/m^3 and 100 to 200 kN/m^2, respectively.

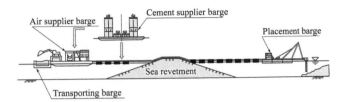

(a) Land reclamation.

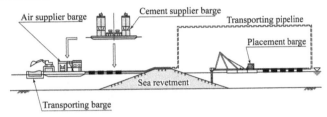

(b) Backfill behind sea revetment.

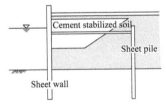

(c) Backfill behind concrete caisson.

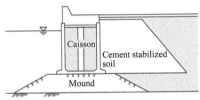

(d) Backfill behind earth retaining structure.

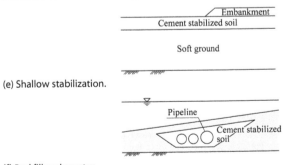

(e) Shallow stabilization.

(f) Backfill underwater.

Figure 1.16 Examples of application of the pneumatic flow mixing method (Ministry of Transport, The Fifth District Port Construction Bureau, 1999).

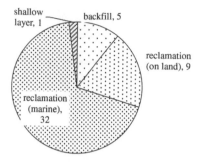

(a) Number of projects.

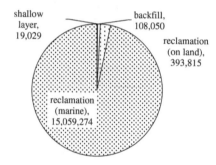

(b) Volume of stabilized soil.

Figure 1.17 Statics of the pneumatic flow mixing method works in Japan.

Figure 1.17 shows summary statistics for use of the pneumatic flow mixing method from 1998 to 2015, which is classified into applications, the number of projects and volumes of stabilized soil. The total number of projects is 47 and the total volume comes to about 15.6 million m³. The application of reclamation in marine areas is dominant; the proportions by number and volume of these projects are about 68.1% and 96.6%.

4 SCOPE OF THE TEXTBOOK

The pneumatic flow mixing method was developed in Japan and put into practice in the late 1990s. The method has been applied to many construction projects since then, in which the total volume of stabilized soil produced by the method from 1998 to 2015 reached about 15.6 million m³. This text book aims to provide the latest State of Practice of the pneumatic flow mixing method to researchers and practitioners based on the researches and experiences accumulated in numerous projects in Japan. The organization of the current textbook is as follows.

Chapter 1 gave an overview of various admixture stabilization techniques. The chapter also showed that a wide range of admixture stabilization techniques, including *in-situ* and *ex-situ* stabilizations, have gained popularity in Japan. Also the development, mechanism and applications of the pneumatic flow mixing method are briefly introduced.

Chapter 2 discusses the influence of various factors on the increase in soil strength brought about by its stabilization with cement. The information compiled in this chapter is basically applicable to all the types of admixture stabilization and useful for evaluating the feasibility of admixture stabilization to a specific soil, in the selection of appropriate binder, and in interpreting the laboratory or field test results.

Chapter 3 describes the engineering properties of stabilized soil mainly stabilized by cement. Facility in transport and placement ability are key factors for the execution of the method, which is influenced by the engineering properties of freshly stabilized soil, soon after mixing. This chapter introduces not only the properties of hardened stabilized soil, the properties of fresh stabilized soil, and the properties of soil that has been subjected to disturbance and compaction.

Chapter 4 describes nine case histories of application of the method in Japan, which cover various types of application, binder and mixing systems; land reclamation; backfill; and shallow layer construction.

Chapter 5 describes the machine system; execution; and quality control and quality assurance during production. The placement technique and process of stabilized soil influences the properties of the field stabilized soil, and the degree of water pollution due to the stabilized soil. This chapter briefly describes the environmental impact of placing stabilized soils in underwater locations, and the countermeasures too. The chapter concentrates on the relevant issues in systems commonly used in Japan.

Chapter 6 describes the geotechnical design procedure currently employed in Japan. The geotechnical design is a way to determine the required quality of stabilized soil and required geometry of stabilized soil ground. This chapter describes the phenomenon of volume change due to stabilization, which is one of the critical design considerations, especially in the case of land reclamation projects on a huge scale.

The Appendix explains the standard laboratory mix test procedure used to test stabilized soils in Japan, with visual examples.

REFERENCES

Akagawa, K. (1980) Gas-liquid two-phase flow. *Mechanical Engineering* 11. p. 15 (in Japanese).
Coastal Development Institute of Technology (2001) *Technical Manual of Pneumatic Flow Mixing Method.* Daikousha Publishers, 127p. (in Japanese).
Coastal Development Institute of Technology (2003) *The Premixing method: Principle, Design and Construction.* A.A.Balkema Publishers. 140p.
Coastal Development Institute of Technology (2008a) *Technical Manual of Lightweight Geomaterial,* revised version. Daikousha Publishers, 370p. (in Japanese).
Coastal Development Institute of Technology (2008b) *Technical Manual of Pneumatic Flow Mixing Method,* revised version. Daikousha Publishers, 188p. (in Japanese).
Coastal Development Institute of Technology (2008c) *Technical Manual of Premixing Method,* revised version. Daikousha Publishers, 215p. (in Japanese).
http://www.penta-ocean.co.jp/english/business/envi/dirt_recycle.html

Kitazume, M. (2004) Construction of a man-made island for Central Japan International Airport by the pneumatic flow mixing method. *Proc. of the 5th International Conference on Ground Improvement Techniques*. pp. 169–176.

Kitazume, M. (2010) Application of Cement Stabilization Methods to Earthquake Disaster Mitigation. *Proc. of the 9th International Symposium on New Technologies for Urban Safety of Mega Cities in Asia*. CD-ROM.

Kitazume, M. & Satoh, T. (2003) Development of Pneumatic Flow Mixing Method and its Application to Central Japan International Airport Construction. *Journal of Ground Improvement*. Vol. 7, No. 3, pp. 139–148.

Kitazume, M. & Satoh, T. (2005) Quality control in Central Japan International Airport Construction. *Journal of Ground Improvement*. Vol. 9, No. 2, pp. 59–66.

Kitazume, M. & Terashi, M. (2013) *The Deep Mixing Method*. CRC Press, Taylor & Francis Group. 410p.

Kitazume, M., Adachi, Y., Ikenoue, N. & Okubo, Y. (2007) Engineering property of treated soil by pneumatic flow mixing method: (Part 3) Relation between the fluidity of treated soil and required air pressure for transport. *Proc. of the 42nd Annual Conference of the Japanese Geotechnical Society*. pp. 607–608 (in Japanese).

Kobayashi, K., Yoshida, G. & Sato, H. (2001) Quality assurance of pneumatic flow mixing method: Snake Mixer Method. *Proc. of the Annual Research Conference, Civil Engineering Research Institute for Cold Region, Ministry of Ministry of Land, Infrastructure, Transport and Tourism*. pp. 393–400 (in Japanese).

Ministry of Land, Infrastructure, Transport & Tourism (2007) *Technical Standards for Port and Harbour Facilities*. Vol. 2, pp. 672–762 (in Japanese).

Ministry of Transport, The Fifth District Port Construction Bureau (1999) *Pneumatic flow mixing method*. Yasuki Publishers. 157p. (in Japanese).

Mizukami, J. & Matsunaga, Y. (2015) Construction of D-Runway at Tokyo International Airport. *Proc. of the 15th Asian Regional Conference on Soil Mechanics and Geotechnical Engineering*. CD-ROM.

Power Blender Method Association (2006) *Technical data on slurry type Power Blender Method*. Power Blender Association. (in Japanese).

Tsuchida, T. & Egashira, K. (2004) *The Lightweight Treated Soil method: New Geomaterials for Soft Ground Engineering in Coastal Areas*. A.A.Balkema Publishers. 120p.

Watanabe, A. (2005) Challenge to soft soil: Pneumatic flow mixing method for dredged soil. *Proc. of the Annual Research Conference, Ministry of Ministry of Land, Infrastructure, Transport and Tourism*. (in Japanese).

Yamagoshi, Y., Akashi, Y., Nakagawa, M., Kanno, H., Tanaka, Y., Tsuji, T., Imamura, T. & Shibuya, T. (2013) Reclamation of the artificial ground made of dredged soil and converter slag by using pipe mixing method. *Journal of Geotechnical Engineering, Japan Society of Civil Engineers*. Vol. 69, No. 2, pp. 952–957 (in Japanese).

Yamauchi, H., Ishiyama, Y. & Oonishi, F. (2011) Study for beneficial use of soft dredged soil: Case history of backfill behind the breakwater at Kushiro Port. *Proc. of the Annual Technical Meeting, Civil Engineering Research Institute for Cold Region*. (in Japanese).

Zen, K., Yamazaki, H., Watanabe, A., Yoshizawa, H. & Tamai, A. (1987) Study on a reclamation method with cement-mixed sandy soils: Fundamental characteristics of treated soils and model tests on the mixing and reclamation. *Technical Note of the Port and Harbour Research Institute*. No. 579, 41p. (in Japanese).

Chapter 2

Factors affecting strength increase

1 INTRODUCTION

The strength increase of cement-stabilized soils is influenced by a number of factors, because the basic strength increase mechanism is closely related to the chemical reaction between soil and binder. The factors can be roughly divided into four categories: I. Characteristics of binder, II. Characteristics and conditions of soil, III. Mixing conditions, and IV. Curing conditions, as shown in Table 2.1 (after Terashi, 1997).

The characteristics of the binder mentioned in Category I strongly affect the strength of the stabilized soil. Therefore, the selection of an appropriate binder is an important issue. There are many types of binder available on the Japanese market (Japan Lime Association, 2009, Japan Cement Association, 2012). The basic mechanisms of admixture stabilization using quicklime or cement were extensively studied by highway engineers many years ago. This is because lime- and cement-stabilized soils have been used as sub-base or subgrade materials in road construction (e.g. Ingles & Metcalf, 1972). The stabilization mechanisms of various binders have been investigated further by geotechnical engineers (e.g. Babasaki et al., 1996a, 1996b). The factors in Category II (characteristics and conditions of the soil) are inherent characteristics of each soil and the way it has been deposited. Thompson (1966) studied the influence of the properties of Illinois soils on the lime reactivity of compacted lime-soil mixtures and concluded that the major influential factors were acidity (potential hydration) and organic matter content of the original soil. Japanese research groups have also performed similar studies on lime- and cement-stabilized soils manufactured without compaction (e.g. Okumura et al., 1974; Kawasaki et al., 1978, 1981, Terashi et al., 1977, 1979, 1980, 1983; Saitoh et al., 1980, 1985; Saitoh, 1988). Their valuable works have provided engineers with good qualitative information. The factors in Category III (mixing conditions) are easily altered and controlled on site during execution based on the judgment of engineers responsible for the execution. The factors in Category IV (curing conditions) can be controlled easily in the laboratory but cannot be controlled at work sites in most cases.

As cement has often been used in the pneumatic flow mixing method, the influence of various factors on the strength of cement-stabilized soil are briefly described in the following section, where the unconfined compressive strength, q_u, of stabilized soil is mainly used as an index representing the stabilization effect. The test specimen for the unconfined compression test is, in principle, prepared in laboratory by the procedure

Table 2.1 Factors affecting the strength increase of stabilized soil (after Terashi, 1997).

I.	Characteristics of binder	1. Type of binder 2. Quality of binder 3. Mixing water and additives
II.	Characteristics and conditions of soil (especially important for clays)	1. Physical, chemical and mineralogical properties of soil 2. Organic content 3. Potential Hydrogen (pH) of pore water 4. Water content
III.	Mixing conditions	1. Quantity of binder 2. Degree of mixing 3. Timing of mixing/re-mixing 4. Placement condition
IV.	Curing conditions	1. Temperature 2. Curing period 3. Humidity 4. Wetting and drying/freezing and thawing, etc. 5. Overburden pressure 6. Soil disturbance/compaction

standardized by the Japanese Geotechnical Society (formerly Japanese Society of Soil Mechanics and Foundation Engineering). The test procedure was originally proposed by Terashi et al. (1980) and welcomed by Japanese researchers and engineers. Essentially the same procedure was adopted by the Japanese Society of Soil Mechanics and Foundation Engineering in 1981 as its Draft Standard JSF: T31-81T in 1982. The draft standard was later officially standardized by the Japanese Society of Soil Mechanics and Foundation Engineering in 1990 and was given minor revisions by the Japanese Geotechnical Society in 2000 and 2009 (Japanese Geotechnical Society, 2000, 2009). The laboratory test procedure is described in the Appendix.

2   MECHANISM OF CEMENT STABILIZATION

In Japan the types of cement used as a binder is usually ordinary Portland cement (OPC) and blast furnace slag cement type B. Ordinary Portland cement is manufactured by adding gypsum to cement clinker and grinding it to powder. Cement clinker is formed by minerals $3CaO.SiO_2$, $2CaO.SiO_3$, $3CaO.Al_2O_3$ and $4CaO.Al_2O_3.Fe_2O_3$. A cement mineral, $3CaO.SiO_2$, for example, reacts with water in the following way to produce cement hydration products.

$$2(3CaO.SiO_2) + 6H_2O = 3CaO.2SiO_2.3H_2O + 3Ca(OH)_2 \qquad (2.1)$$

During the hydration of cement, calcium hydroxide, $Ca(OH)_2$, is released. The cement hydration product has high strength, which increases as it ages, while calcium hydroxide contributes to the pozzolanic reaction. Blast furnace slag cement is a mixture of ordinary Portland cement and blast furnace slag. Finely powdered blast furnace slag does not react with water but has the potential to produce pozzolanic reaction

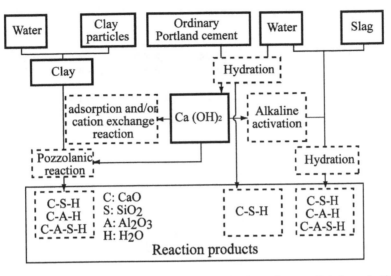

Figure 2.1 Chemical reactions between clay, cement, slag and water (Saitoh et al., 1985).

products under high alkaline conditions. In blast furnace slag cement, silicon dioxide, SiO_2, and aluminium oxide, Al_2O_3, contained in the slag are actively released by the stimulus of large quantities of Ca^{2+} and SO_4^{2-} released from the cement, so that fine hydration products abounding in silicates are formed, rather than cement hydration products, and the long term strength is enhanced. The rather complicated mechanism of cement stabilization is simplified and schematically shown in Figure 2.1 for the chemical reactions between clay, pore water, cement and slag (Saitoh et al., 1985).

3 INFLUENCE OF VARIOUS FACTORS ON THE STABILIZATION EFFECT

3.1 Influence of the characteristics of the binder

3.1.1 Chemical composition of the binder

In Japan, ordinary Portland cement (OPC) and blast furnace slag cement type B have often been used as a binder for stabilizing clay and sand, whose chemical components are specified by Japanese Industrial Standard (Japanese Industrial Standard, 2006, 2009) as tabulated in Table 2.2. In addition to the two types of cement, several cement-based special binders have been developed and are available on the Japanese market as shown in Table 2.3 (Japan Cement Association, 2012).

The cement-based special binders are specially manufactured for the specific purpose of stabilizing soil or similar material by increasing certain constituents of ordinary Portland cement. This is done by adjusting its Blaine specific surface area or by adding ingredients effective for particular soil types. These are actually a mixture of cement

Table 2.2   Chemical components of Japanese cements (Japanese Industrial Standard, 2006, 2009).

	CaO (%)	SiO$_2$ (%)	Al$_2$O$_4$ (%)	Fe$_2$O$_4$ (%)	SO$_4$ (%)	Others
Ordinary Portland cement	64–65	20–24	4.8–5.8	2.5–4.6	1.5–2.4	MgO,
High-early-strength Portland cement	64–66	20–22	4.0–5.2	2.4–4.4	2.5–4.4	Na$_2$O, K$_2$O,
Blast furnace slag cement type B	52–58	24–27	7.0–9.5	1.6–2.5	1.2–2.6	MnO, P$_2$O$_5$

Table 2.3   Cement-based special binders (Japan Cement Association, 2012).

Type	Characteristics
For soft soils	Appropriate for soft soils with a high water content, e.g. sand, silt, clay and volcanic soil
For problematic soils	To reduce leaching of hexavalent chromium (chromium VI) from stabilized soil
For organic soils	Appropriate for highly organic soils, e.g. humus, organic soil, sludge

as a mother material and gypsum, a micropowder of slag, alumina or fly ash. The chemical components of the cement-based special binders are not specified by the Japanese Industrial Standard and are not published as proprietary information by cement manufacturers. As shown in Table 2.3, the cement-based special binders are designed for high water content soil, high organic content soil and for reducing the leaching of Cr^{6+} from stabilized soil. The stabilization effect in organic soil is said to be affected by the composite ratio, $(SiO_2 + Al_2O_3)$ against CaO, of the constituent elements in cement and cement-based special binders (Hayashi et al., 1989).

Other than those special binders, 'delayed stabilizing' or 'long-term strength control' type binders are also available on the Japanese market, by which the rate of strength increase can be controlled. They are obtained by adjusting the quantities of ingredients such as gypsum or lime. These binders react slowly with soil and exhibit smaller strength in the short term, but result in sufficiently high strengths in the long term in comparison with ordinary Portland cement or blast furnace slag cement type B (Kiyota et al., 2003).

An example of the effects of the chemical compounds calcium oxide, CaO, sulphur trioxide, SO_3, and aluminium oxide, Al_2O_3, on the strength is shown in Figure 2.2 (Japan Cement Association, 2009). In the test, dredged clay (w_L of 60.7%, w_P of 29.1% and I_p of 31) was stabilized with mixture of several types of cement and cement-based special binders so that the effects of chemical compounds could be highlighted. After four weeks curing, the stabilized soils were subjected to unconfined compression test. The unconfined compressive strength, q_u is compared with the content of chemical compounds in the binder. In the effect of CaO, Figure 2.2(a), the strength remains almost constant irrespective of the amount of CaO as long as the amount of binder is about 80 kg/m^3. When the amount of binder is increased to 140 and 200 kg/m^3,

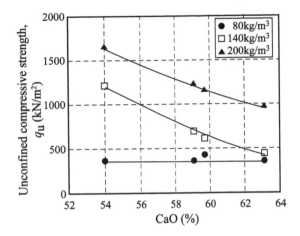

(a) Effect of CaO content.

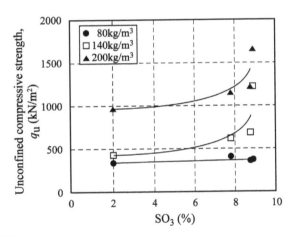

(b) Effect of SO₃ content.

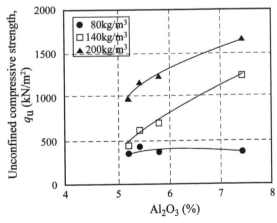

(c) Effect of Al₂O₃ content.

Figure 2.2 The effects of chemical compounds on the strength of cement-stabilized soil (Japan Cement Association, 2009).

however, the strength decreases with the content of CaO. In the effect of SO_3, Figure 2.2(b), the strength is almost constant irrespective of the amount of binder as long as the amount of SO_3 remains lower than about 8%. However, when the amount of SO_3 becomes about 9%, the strength rapidly increases. In the effect of Al_2O_3, Figure 2.2(c), the strength remains almost constant irrespective of the amount of Al_2O_3 as long as the amount of binder is about $80\,kg/m^3$. When the amount of binder is increased to 140 and $200\,kg/m^3$, however, the strength increases almost linearly with the Al_2O_3 content.

3.1.2 Type of binder

Figure 2.3 shows the influence of the type of cement on the strength of stabilized soil in which ordinary Portland cement and blast furnace slag cement type B were compared at the curing period, with a t_c of 28 days to 5 years (Saitoh, 1988). The tests were conducted on two different sea bottom sediments: the Yokohama Port clay (w_L of 95.4%, w_P of 42.4% and w_i of 97.9%) and the Osaka Port clay (w_L of 79.4%, w_P of 40.2% and w_i of 94.9%). For each clay, three different amounts of cement, α of 100 to $300\,kg/m^3$, were mixed. The cement factor, α, is defined as a dry weight of cement added to $1\,m^3$ of original soil. The horizontal axes of the figures show the curing period, t_c. The vertical axis of the upper figures for each clay is the unconfined compressive strength, q_u, of the stabilized soil, while the vertical axes of the lower figures is the normalized unconfined compressive strength at an arbitrary curing period, t_c, by the 28 days' strength: q_{utc}/q_{u28}. In the case of the Yokohama Port clay, which exhibits high pozzolanic reactivity, ordinary Portland cement is much more effective than blast furnace slag cement type B. However, in the case of the Osaka Port clay, with its lower pozzolanic reactivity than the Yokohama Port clay, blast furnace slag cement type B is much more effective. These test results suggest that the appropriate selection of the type of cement can be made if the pozzolanic reactivity of soil is evaluated beforehand. It is interesting to see the q_{utc}/q_{u28} is higher for blast furnace slag cement type B than for ordinary Portland cement, irrespective of the soil type.

Figure 2.4 shows the influence of various cement-based special binders on the strength of various types of organic soil (Coastal Development Institute of Technology, 2008a). The physical and chemical properties of the soils are tabulated in Table 2.4 (Coastal Development Institute of Technology, 2008a). The letters along the horizontal axes of the figures represent the types of binder, while the chemical components of some binders are shown in Table 2.5 (Coastal Development Institute of Technology, 2008a). The figures show that cement-based special binders are effective for organic soils in general, but that the most effective binder for a particular soil is not always the best binder for the other types of organic soil. For these problematic soils, appropriate selection of binder by a laboratory mix test is important. A similar phenomenon on the strength of stabilized organic soils will be shown in Figure 2.7.

3.1.3 Type of mixing water

Table 2.6 shows the influence of the type of mixing water for preparing binder slurry on the strength of stabilized soil, where clay excavated at Tokyo Port (w_L of 94.1% and w_P of 45.8%) was stabilized with ordinary Portland cement (Kawasaki et al., 1978).

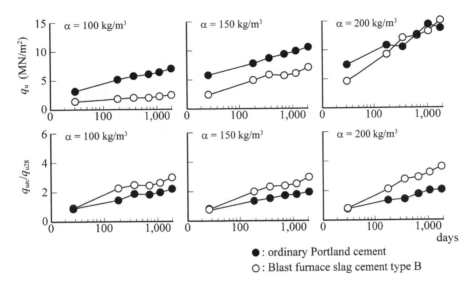

(a) Influence of binder type on the Yokohama Port clay.

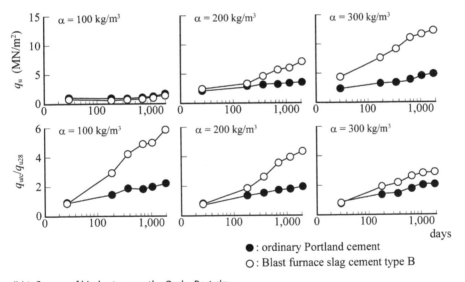

(b) Influence of binder type on the Osaka Port clay.

Figure 2.3 The influence of cement type on the unconfined compressive strength of cement-stabilized marine clays (Saitoh, 1988).

The cement slurry was prepared by two types of water: tap water and seawater obtained at Tokyo Port. The table shows that the strength of the stabilized soil with tap water is slightly smaller than that with seawater, but the difference is negligibly small from a practical point of view.

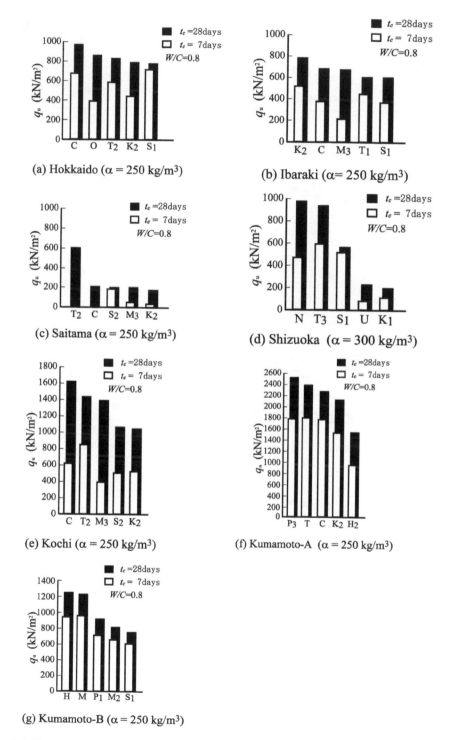

Figure 2.4 The unconfined compressive strength of organic soils stabilized with cement-based special binders (Coastal Development Institute of Technology, 2008a).

Table 2.4 Physical and chemical properties of soils (Coastal Development Institute of Technology, 2008a).

Depth (m)	Hokkaido −0.5 to −1.0	Ibaraki −0.5 to −1.0	Saitama −0.5 to −1.0	Shizuoka −4.0 to −4.0	Kochi −1.0 to −1.5	Kumamoto-A −5.0 to −7.5	Kumamoto-B −0.5 to −1.0
Grain size distribution							
gravel (%)	—	—	—	—	0.0	0	0.0
sand (%)	—	—	—	—	0.0	0	2.9
silt (%)	—	—	—	—	71.8	40.5	42.0
clay (%)	—	—	—	—	28.2	59.5	55.1
Consistency limits							
liquid limit, w_L (%)	—	251.2	—	—	271.6	174.8	181.4
plastic limit, w_P (%)	—	92.7	—	—	69.1	76.2	47.4
plasticity index, I_p	—	158.5	—	—	202.5	97.6	144.0
Particle density (g/cm³)	1.969	1.688	2.099	1.700	2.249	2.279	1.572
Natural condition							
water content (%)	492	246	940	840	295	156.4	159
density, ρ_c (g/cm³)	1.11	1.16	1.04	1.045	1.14	1.400	1.26
Chemical properties							
ignition loss (%)	55.2	47.7	67.4	70.5	24.8	22.2	24.0
dichromate (%)	42.4	25.2	59.0	—	17.6	—	11.5
humus content (%)	8.1	15.2	28.6	17.2	4.1	—	7.4
pH	4.9	4.7	4.5	—	4.0	6.7	5.0

Table 2.5 Chemical components of binders (Coastal Development Institute of Technology, 2008a).

Binder	SiO_2	Al_2O_3	Fe_2O_2	CaO	MgO	SO_3	Na_2O	K_2O
C	21.5	7.8	1.6	51.7	2.5	9.4	0.2	0.4
H	20.8	8.3	2.0	53.0	3.1	9.7	0.3	0.3
M	17.3	4.9	2.5	59.9	1.8	8.4	0.1	0.1
N	19.8	7.3	1.8	53.0	2.6	12.9	0.1	0.1
O	17.6	4.5	2.9	57.8	1.4	11.3	0.4	0.5

Table 2.6 Influence of type of water used to prepare the cement slurry on strength of stabilized soil (Kawasaki et al., 1978).

Initial water (%) content, w_i	Cement content, aw (%)	Curing period (day)	Unconfined compressive strength, q_u (kN/m^2)		Strength ratio q_{u_tap}/q_{u_sea}
			tap water	seawater	
79.9	13.1	7	2400	2640	0.91
	13.1	28	3500	3700	0.95
85.1	13.1	7	2080	2090	0.99
	13.1	28	3090	2980	1.04

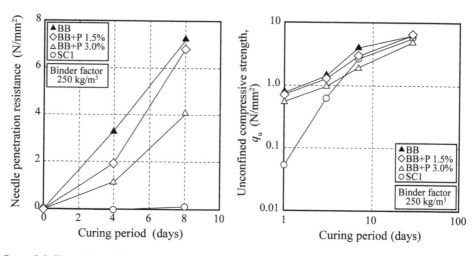

Figure 2.5 The effect of the stabilizing agent on the strength of stabilized soil (Kiyota et al., 2003).

3.1.4 Type of additives

Figure 2.5 shows the influence of the type of slow-setting special cement on the strength of stabilized soil (Kiyota et al., 2003). In the tests, clay excavated in Tokyo (w_i of 51.3%) was stabilized with either a slow-setting special cement (SC1), blast furnace slag cement type B (BB), or mixture of blast furnace slag cement type B and a slow-setting special cement (BB+P 1.5%, BB+P 3.0%). The stabilized soils were subjected to the needle penetration tests (JIS A 6204) within one day's curing and an unconfined compression test after one day's curing. Figure 2.5(a) shows that the needle penetration resistance is quite small in the case of the SC1, while the resistances of the BB+P 1.5% and BB+P 3.0% are a little smaller than the BB-stabilized soil. The unconfined compressive strength of SC1 increases very rapidly after one day and reaches almost the same q_u values as BB-stabilized soil at 28 days' curing, as shown in Figure 2.5(b).

3.2   Influence of the characteristics and conditions of soil

3.2.1   Soil type

In order to investigate the influential factors on cement stabilization, Babasaki et al. (1996a, 1996b) collected 231 test results on soils taken from 69 locations in Japan from the 14 papers published during 1981 to 1992 in Japan. For deducing the influence of soil type from the test data conducted in different laboratories, the other factors listed in Table 2.1 were kept constant. Regarding the characteristics of binder, the test data on ordinary Portland cement and blast furnace slag cement type B were compared, while the mixing and curing conditions were the same for all the tests. Figure 2.6 compares the binder content, aw, and unconfined compressive strength at 28 days' curing, q_u, for various soils. Even for the same binder content of aw, the q_u varies considerably according to the type of soil. The strength of a particular cement-stabilized soil increases with the amount of cement, as later shown in Figures 2.17 and 2.18. The large variation of strength found in Figure 2.6 clearly reveals that the strength gain by cement stabilization depends heavily upon the type and properties of soil.

The influence of soil type on the unconfined compressive strength, q_u, is also shown in Figure 2.7, in which a total of 21 different soils with various physical and chemical properties were stabilized by ordinary Portland cement with a cement content, aw, of

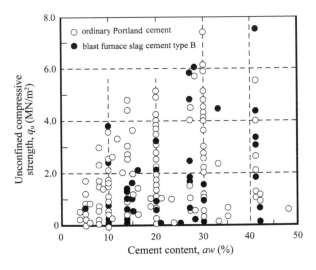

Figure 2.6 The relationship between unconfined compressive strength, q_u, and cement content, aw (Babasaki et al., 1996a, 1996b).

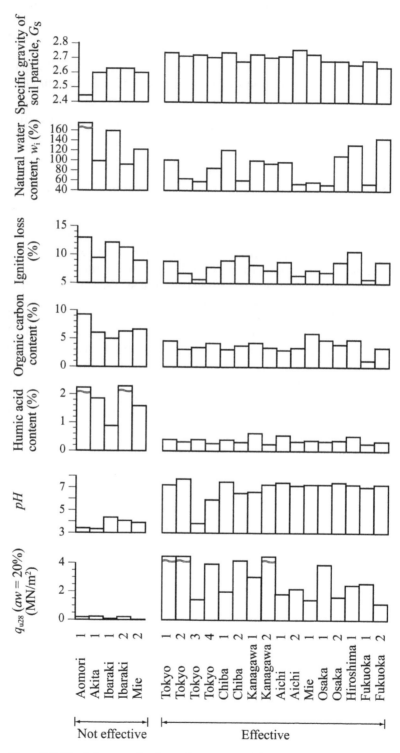

Figure 2.7 The influence of soil type in cement stabilization (Niina et al., 1981).

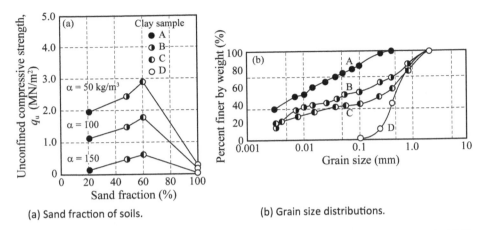

(a) Sand fraction of soils.   (b) Grain size distributions.

Figure 2.8 The influence of the grain size distribution in cement stabilization (Niina et al., 1977).

20% (Niina et al., 1981). The figure shows that the humic acid content and pH of soil are the most dominant factors influencing the strength gain.

3.2.2 Grain size distribution

Figure 2.8(a) shows the influence of grain size distribution of soil on the unconfined compressive strength, q_u, of cement-stabilized soil (Niina et al., 1977). In addition to the two natural soils, two artificial soils, B and C, were prepared by mixing the two natural soils, the Shinagawa alluvial clay (w_L of 62.6% and w_P of 24.1%) named A, and the Ooigawa sand, named D, whose grain size distributions are shown in Figure 2.8(b). These soils were stabilized with ordinary Portland cement with three amounts of cement factor, α, and the unconfined compression tests were carried out on the stabilized soils after 28 days' curing. The unconfined compressive strength, q_u, is influenced by the grain size distribution and the highest stabilization effect can be achieved at around 60% of sand fraction, irrespective of the amount of cement. However, this is not always true, as the test result on the other type of soil shows that the strength of stabilized soil is quite different even with almost the same grain size distribution, which may be influenced by its consistency property (Okano et al., 2012).

3.2.3 Humic acid

Figure 2.9 shows the influence of humic acid content on the unconfined compressive strength of cement-stabilized soil (Okada et al., 1983). Artificial soil samples were prepared by mixing various amount of humic acid with the kaolin clay (w_L of 50.6%), in which three kinds of humic acid extracted from Japanese clays and commercially available humic acid were mixed. These artificial soils having the same initial water content of 60% were stabilized with cement content, aw, of 5% ordinary Portland cement, while the humic acid content and the dry weight of the soil were changed. The figure clearly shows that the influence of the humic acid on the strength depends on its characteristics: the acid extracted from the Negina River clay has a negligible

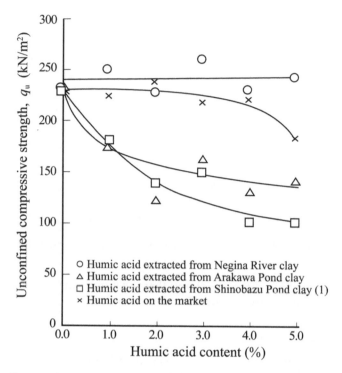

Figure 2.9 The influence of humic acid content on the unconfined compressive strength of stabilized clays (Okada et al., 1983).

influence on the strength, while the acid extracted from the Shinobazu Pond clay gives a considerably large influence on the strength.

Figure 2.10 also shows the influence of the humic acid content of soil on unconfined compressive strength (Miki et al., 1984). Artificial soil samples were prepared by adding various amounts of humic acid extracted from the clay at Arakawa Pond to the kaolin clay (w_L of 50.6%). In the tests, the artificial soils were stabilized with nine types of binder whose chemical compositions are shown in Figure 2.10(a). Figure 2.10(b) shows that the unconfined compressive strength, q_u, is highly dependent upon the binder but decreases considerably with the humic acid content, irrespective of the type of binder. The strength decreases to about one third when the humic acid content is increased to about 5%.

3.2.4   Organic content

3.2.4.1   Ignition loss

Figure 2.11 shows the relationship between ignition loss and the unconfined compressive strength, q_u, of stabilized soils (Babasaki et al., 1996a, 1996b). When the ignition loss is smaller than 15%, a relatively large strength can be generally achieved by mixing a certain amount of cement. For soils with ignition loss exceeding 15%, on the other hand, the unconfined compressive strength, q_u, remains low even with the cement content, *aw*, exceeding 20%, which means that large strength cannot be achieved with use of a practical amount of cement. The soils within the circled half-tone dot mesh

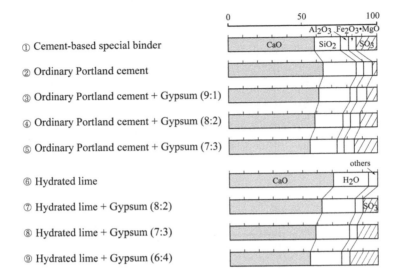

(a) Chemical composition of binders.

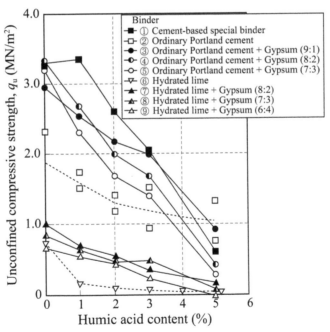

(b) Unconfined compressive strength, q_u.

Figure 2.10 The influence of humic acid content on the unconfined compressive strength of stabilized clays (Miki et al., 1984).

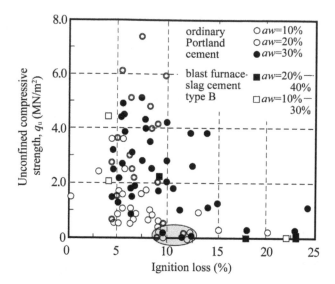

Figure 2.11 The relationship between unconfined compressive strength, q_u, and ignition loss of cement stabilized soils (Babasaki et al., 1996a, 1996b).

in the figure do not show an increase in strength, despite the increase in the cement content. The ignition loss of the soils is lower than 15% but the proportion of humus in the soil exceeds 0.9%, higher than that of the original soils, which may cause the quite low strength. Although there are some exceptions, ignition loss is one of the key indices with which to evaluate the stabilization effect.

3.2.4.2 Potential hydrogen (pH) of soil

Figure 2.12 shows the relationship between the potential hydrogen, pH, of original soil and unconfined compressive strength, q_u, (Babasaki et al., 1996a, 1996b). As the figure shows, most of soils with a pH value lower than 5 show a smaller strength increase than those with a pH value higher than 5 for the same binder content. Although there are exceptions in which the stabilization effect is not low, even with a low pH value, the pH value is one of the key indices to evaluate the stabilization effect.

The relationship between the pH of original soil and the unconfined compressive strength, q_u, of stabilized soil is proposed as Figure 2.13 and Equation 2.2 (Nakamura et al., 1980).

$$q_u = 0.0325 \times F - 1.625$$

(2.2)

$$\left. \begin{array}{ll} F = Wc/(9 - pH) & for \, pH < 8 \\ F = Wc & for \, pH > 8 \end{array} \right\}$$

where
F: parameter (kg/m^3)
pH: potential hydrogen
q_u: unconfined compressive strength (MN/m^2)
Wc: cement factor (kg/m^3)

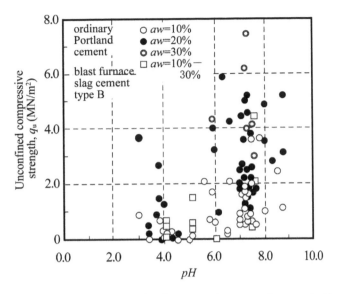

Figure 2.12 The relationship between unconfined compressive strength, q_u, and pH of original soils (Babasaki et al., 1996a, 1996b).

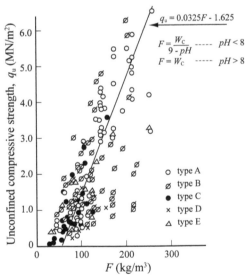

Type	Wet unit weight, γ_t (g/cm³)	Natural water content, w_n (%)	Liquid limit w_L (%)	Plastic limit w_p (%)	Grain size distribution (%)			pH (H₂O)	Ignition loss
					Sand, gravel fraction	Silt fraction	Clay fraction		
A	1.38-1.76	55-144	51-121	2-18	30-47	33-50	29-52	8.1-8.7	2.0-7.0
B	1.28-1.70	38-160	27-204	3-42	18-76	27-70	16-54	7.3-8.9	3.8-12.9
C	1.50-1.76	42-86	49-110	3-43	22-66	36-54	12-55	5.5-7.9	4.0-12.0
D	1.10-1.40	114-740	–	–	–	–	–	5.5-6.0	19.0-64.0
E	1.49-1.97	25-56	–	49-86	–	9-35	2-2	5.4-9.3	3.3-15.2

Figure 2.13 The effects of pH of soil on cement-stabilized soil (Nakamura et al., 1980).

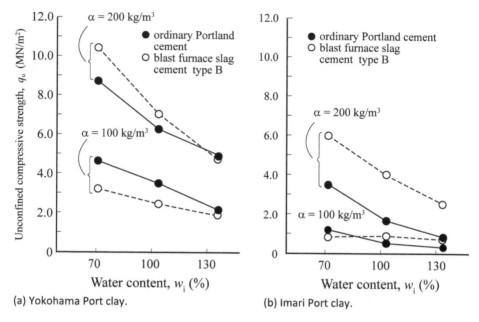

Figure 2.14 The influence of initial water content on the strength of two cement-stabilized clays (t_c of 91 days) (Saitoh, 1988).

3.2.5 Water content

The influence of the initial water content of the soil on unconfined compressive strength, q_u, of stabilized soil is shown in Figure 2.14 (Saitoh, 1988). In the tests, two kinds of marine clay (the Yokohama Port clay, w_L of 95.4%, w_P of 32.3% and w_n of 97.9%; and the Imari Port clay, w_L of 70.4%, w_P of 24.2% and w_n of 83.3%) were stabilized with either ordinary Portland cement or blast furnace slag cement type B. The Yokohama Port clay has high pozzolanic activity, while the Imari Port clay has low pozzolanic activity. The magnitude of strength gain is large for the high pozzolanic activity soil but not for the low pozzolanic activity soil. The unconfined compressive strength decreases almost linearly with the initial water content, irrespective of the pozzolanic activity of soil and cement type.

Figure 2.15 shows the relationship between the total water content, w_t, defined by the total weight of water, including pore water and mixing water, and the dry weight of soil and the q_u of stabilized soil with various cement contents, aw, of 10, 20, 30 and 35% (Babasaki et al., 1996a, 1996b). The figure shows that the strength of stabilized soil decreases rapidly with the total water content. Soils with a total water content, w_t, higher than 200% cannot achieve large gains in strength even with a large amount of cement. There are some exceptions where it is difficult to improve the soils' physical and mechanical properties even when their water content is lower than 200%. These soils usually contain a high amount of organic material, or are acidic soils with a low pH value; it is preferable to stabilize these soils with the cement-based special binders as already shown in Table 2.3.

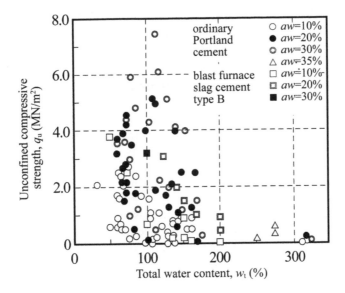

Figure 2.15 The relationship between unconfined compressive strength, q_u, and total water content, w_t of soil (Babasaki et al., 1996a, 1996b).

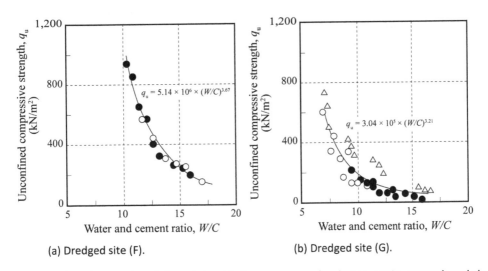

(a) Dredged site (F).   (b) Dredged site (G).

Figure 2.16 Typical examples of the relationship between unconfined compressive strength and the W/C ratio of soil (Kitazume & Satoh, 2003).

It has been well known that unconfined compressive strength, q_u, has a close relationship to the W/C ratio, in which the W/C ratio is defined as the total water contained in both soil and cement slurry against the dry weight of the cement. Figure 2.16 shows typical examples of the relationship, in which two types of soil (w_L of 78.3% and 55.3%, respectively) excavated at Nagoya Port were prepared to have various initial water contents and stabilized with blast furnace slag cement type B (Kitazume & Satoh, 2003). The figure shows that the q_u is almost inversely proportional to the W/C, irrespective of the soil type. Similar relationships were obtained on various soils in laboratory mix tests and field tests (Ministry of Transport, The Fifth District Port Construction Bureau, 1999).

3.3   Influence of the mixing conditions

3.3.1   Quantity of cement

Figure 2.17 shows the influence of the cement content, aw, on the unconfined compressive strength, q_u, in which the Fukuyama Port clay (w_L of 97.6% and w_P of 33.8%), having an initial water content of $1.5 \times w_L$ (146%), was stabilized with ordinary Portland cement, and was subjected to an unconfined compression test at three curing periods (Udaka et al., 2013). The unconfined compressive strength increases almost linearly with the amount of cement. The figure also shows that a minimum amount of cement – about 5% – is necessary to obtain a certain magnitude of strength gain for this particular soil, irrespective of the curing period. A similar phenomenon was also found in the previous research (Terashi et al., 1980).

A similar phenomenon for organic soils is shown in Figure 2.18 (Babasaki et al., 1980). In the tests, four soils with various initial water contents were stabilized with ordinary Portland cement. The unconfined compressive strength is relatively small in the organic soils, but it increases with the cement factor. The figure clearly shows that the minimum cement factor of around $50\,kg/m^3$ is necessary to achieve an appreciable strength increase, which corresponds the cement content of about 7% for G_s of 2.65 and w of 100%. Regarding the minimum amount of cement to obtain an

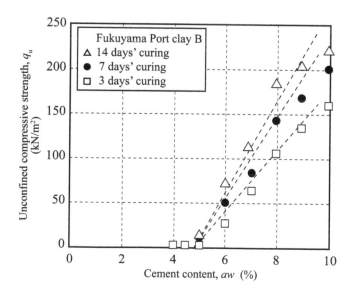

Figure 2.17 The influence of the amount of cement on unconfined compressive strength (Udaka et al., 2013).

stabilization effect, Udaka et al. (2013) investigated five types of soil and concluded that it is about 3.0 to 7.5% cement content, which is close to the test results shown in Figure 2.17.

3.3.2 Mixing time

Figure 2.19 shows the relationship between mixing time and the unconfined compressive strength, q_u, in laboratory mix tests (Nakamura et al., 1982). The laboratory mix tests were conducted as the same manner as the standardized procedure (Japanese Society of Soil Mechanics and Foundation Engineering, 1990) except for the mixing time. In the tests, the Narashino clay (w_i of 68%) was stabilized with ordinary Portland cement in either dry form or slurry form with a water to cement ratio, W/C of 100%. The unconfined compressive strength decreases with decreasing mixing time. The figure also shows that the strength deviation increases with decreasing mixing time. According to this phenomenon and similar test results, the Japanese Society of Soil Mechanics and Foundation Engineering recommends that the mixing time should be about 10 minutes in a laboratory mix test where large strength with small deviation can be sought

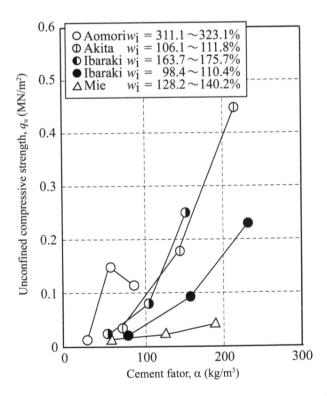

Figure 2.18 The influence of the amount of cement on the strength of cement-stabilized organic soils (Babasaki et al., 1980).

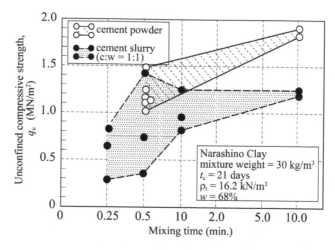

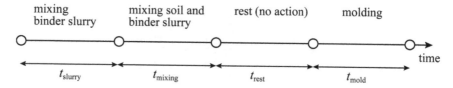

Figure 2.19 The influence of mixing time on the strength and deviation of cement-stabilized soil (Nakamura et al., 1982).

Figure 2.20 Process chart of mixing and molding of cement-stabilized soil.

3.3.3 Time and duration of mixing, and molding process

Figure 2.20 indicates the time sequence of the mixing and molding process in laboratory mix test. In the mix test, a test specimen is usually manufactured by the following steps: 1) disaggregation and homogenization of the original soil, 2) preparation of the cement slurry at a prescribed water-to-cement ratio, W/C, which takes t_{slurry}, 3) mixing soil and cement slurry to prepare a uniform soil–cement mixture, t_{mixing}, 4) rest time before molding, t_{rest}, 5) filling a mold with the soil–cement mixture within t_{mold}. Chemical reactions between cement and water start at steps 2 and 3. As the chemical reactions proceed with time, the time duration in steps 2 to 4 may influence the q_u of stabilized soil. For example, if the time for mixing the cement slurry and soil and/or the time until molding is unnecessarily long, the chemical reaction products at steps 2 to 4 may be broken during the molding procedure at step 5. Also anticipated is that the change of fluidity of the soil–cement mixture may result in difficulty in molding. The time duration of steps 2 to 4 is shown in Figure 2.20.

Although the time for mixing soil and binder slurry is not clearly specified in the Japanese standard, 10 minutes' mixing is the de facto standard in Japan

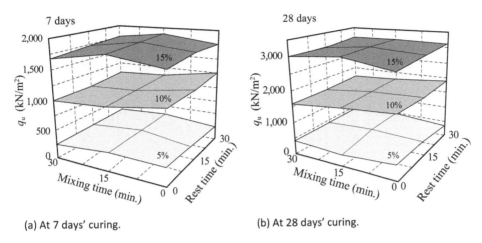

(a) At 7 days' curing.

(b) At 28 days' curing.

Figure 2.21 The influence of cement–slurry mixing time, and rest time after clay–cement mixing, on strength (Kitazume & Nishimura, 2009).

(Japanese Geotechnical Society, 2009). Other time durations are considered to vary considerably from laboratory to laboratory, depending on the number of laboratory technicians and number of specimens prepared from a batch of soil–binder mixture. Any delay in the test procedure may cause a deterioration of the stabilized soil specimens' properties.

Figure 2.21 shows the effects of the mixing time, t_{mixing}, of cement slurry, and the rest time, t_{rest}, on the strength of stabilized soil (Kitazume & Nishimura, 2009). The rest time is defined as the time period between the end of mixing and the start of molding. In the tests, the Kawasaki clay (w_L of 54.1%, w_P of 24.0% and w_i of 65.0%) was stabilized with ordinary Portland cement with a W/C ratio of 100%, in which the cement content, aw, was changed to 5, 10 and 15%. The case $t_{slurry} = 0$ corresponds to the situation where the cement and water are simultaneously added to the soil, or the powder form of cement is added. The unconfined compressive strengths, q_u, measured at 7 and 28 days' curing are shown in Figures 2.21(a) and 2.21(b), respectively. The standard deviation of q_u in each condition (three tests) was 2.6 to 2.9% on average. The test results indicate that the time after mixing cement and water, t_{slurry}, and the time after mixing the soil and cement slurry, t_{rest}, have little influence on the q_u. The unit weight of specimen exhibits little variability, being correlated more to the initial water content of the batches.

Figure 2.22 shows additional test results with extended rest times after mixing, t_{rest}, to identify the limit beyond which the soundness of specimen preparation is compromised (Kitazume & Nishimura, 2009). The test results reveal it is when the t_{rest} exceeds 40 minutes that the specimen quality starts being affected by the soil – cement's reduced fluidity, and hence by the difficulty in 'compacting' through tapping actions, which is the process specified by the Japanese Society of Soil Mechanics and Foundation Engineering Standard (Japanese Society of Soil Mechanics and Foundation Engineering, 1990). Longer t_{rest} resulted in inclusions of numerous voids in the completed specimens, and lower unit weight, which is closely related to q_u.

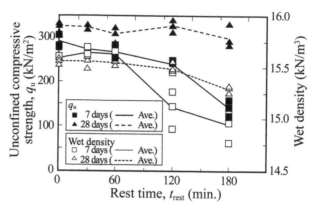

(a) Cement content of 5%.

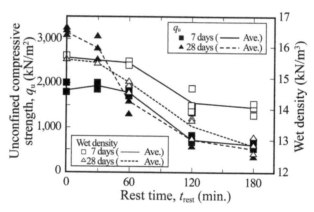

(b) Cement content of 15%.

Figure 2.22 The influence of rest time after clay–cement mixing on the strength (Kitazume & Nishimura, 2009).

3.4 Influence of the curing conditions

3.4.1 Hydration

Figure 2.23 shows the strength increase of cement-stabilized soil soon after the mixing (Watabe et al., 2001). In the tests, the Tachibana Bay clay (w_L of 40.8% and w_P of 20.8%) was prepared to have an initial water content of $1.6 \times w_L$ (65.3%) and stabilized with ordinary Portland cement with the cement factor, α, of 50 kg/m^3. Soon after the mixing, the shear strength, τ_v, was measured with a hand vane shear apparatus at several elapsed times up to 200 minutes. The shear strength of the original soil with a water content of $1.6 \times w_L$ was also measured and plotted in the figure. The shear strength of the stabilized soil just after mixing is larger than the original soil, which is probably due to the reduction in fluidity by adding cement powder. The shear strength,

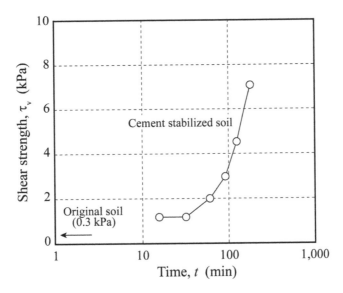

Figure 2.23 Strength increase in cement-stabilized soil soon after cement mixing (Watabe et al., 2001).

τ_{v}, remains a small value within about 30 minutes but increases rapidly after that due to the progress of the cement hydration effect.

Figure 2.24 shows one of comprehensive test results on the strength gain soon after stabilization, in which three kinds of soil were prepared to have several initial water contents, and stabilized with ordinary Portland cement with several cement contents (Tsuchida et al., 2013). The shear strength, s_u, of the stabilized soil was measured with a hand vane apparatus for soils with small strength, or by an unconfined compression test, q_u, for soils with a relatively large strength. Figure 2.24 shows one of the test results on stabilized Mizushima Port clay (w_L of 65.3% and w_P of 15.5%). The cement content, c^*, in the figure is defined by the dry weight of cement against the dry weight of the stabilized soil, instead of the dry weight of cement against the dry weight of the original soil. The figure shows the shear strength increases almost linearly with the curing period in a double logarithm scale up to about 72 hours and then the strength still increases with the curing period, but with lower increment ratio, irrespective of the initial water content and the cement content c^*. A similar phenomenon is also found in the other two soils, the Hibiki clay (w_L of 61.2% and w_P of 20.7%) and the Moji Port clay (w_L of 89.5% and w_P of 29.3%).

3.4.2 Curing period

Figure 2.25 shows the strength increase of cement-stabilized soil with the curing period (Sasayama et al., 2011). In the tests, two soils excavated at Nagoya Bay (w_L of 52.5% and w_P of 25.1% for Soil C, and w_L of 48.7% and w_P of 29.6% for Soil S) were stabilized with ordinary Portland cement with cement factor, C, of 30, 50, 70

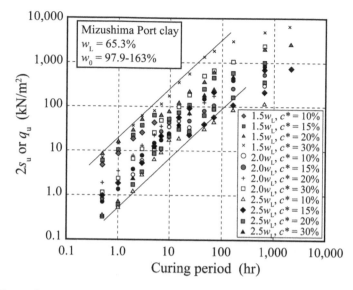

Figure 2.24   Effects of mixing condition on strength increase soon after cement mixing (Tsuchida et al., 2013).

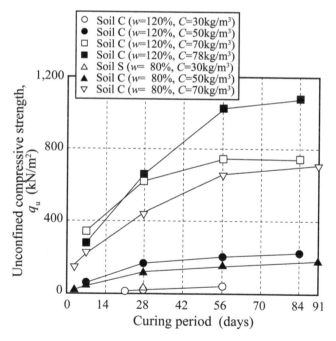

Figure 2.25   Effect of mixing condition on strength increase with curing period (Sasayama et al., 2011).

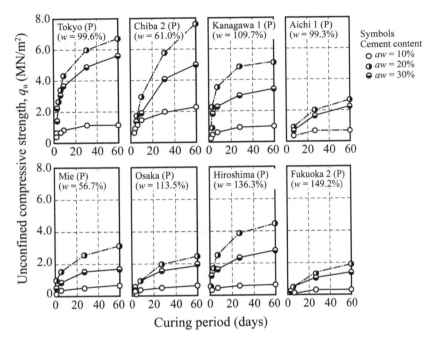

Figure 2.26 Influence of soil type to strength increase with curing period (Kawasaki et al., 1981).

and 78 kg/m³. The unconfined compressive strength, q_u, increases with the curing period irrespective of the soil type and initial water content, and the strength increase ratio with curing period is more dominant for the stabilized soil with a large amount of cement. Similar test results were obtained for soils stabilized with ordinary Portland cement or blast furnace slag cement type B (Saitoh, 1988).

Figure 2.26 shows the strength increase of cement-stabilized soils with the curing period (Kawasaki et al., 1981). In the tests, soils excavated in eight areas were stabilized with ordinary Portland cement with *aw* of 10, 20 and 30%. The unconfined compressive strength, q_u, increases with the curing period irrespective of the soil type and the cement content, and the strength increase ratio with time is more dominant for the stabilized soil with a large amount of cement.

The relationships between the strength of stabilized soils at two different curing periods have been studied. Figures 2.27(a) and 2.27(b) show two typical examples of the relationship for organic soils and cohesive soils, respectively (Cement Deep Mixing Method Association, 1999). In Figure 2.27(a), the strength ratio, q_{u28}/q_{u7}, ranges from 1 to 4 with a mean value of 1.44 for the stabilized organic soils. For the cohesive soils (Figure 2.27(b)), on the other hand, the mean value of q_{u28}/q_{u7} is 1.57. A similar strength ratio, q_{u28}/q_{u7}, of 1.4 to 2.3; q_{u91}/q_{u7} of 1.8 to 5.9; and q_{u91}/q_{u28} of 1.2 to 2.1 for the clay and sand were reported by Saitoh (1988). The strength ratio, q_{u28}/q_{u7}, depends on the soil type, and the type and amount of binder.

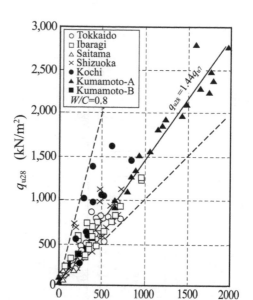

(a) Organic soils.

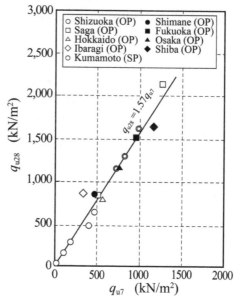

(b) Cohesive soils.

Figure 2.27 The relationship between unconfined compressive strength at 28 days' curing and that at 7 days' curing (Cement Deep Mixing Method Association, 1999).

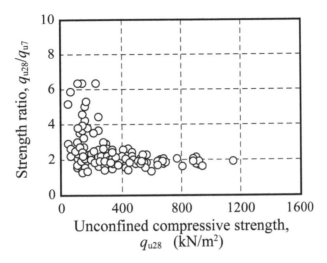

Figure 2.28 The relationship between the unconfined compressive strength and the strength ratio (Coastal Development Institute of Technology, 2008b).

Figure 2.28 shows the relationship between the strength ratio, q_{u28}/q_{u7}, and the q_{u28} on the laboratory stabilized clays with blast furnace slag cement type B (Coastal Development Institute of Technology, 2008b). The figure shows the strength ratio ranges from around 1.5 to 6.5 as far as the q_{u28} is smaller than about 400 kN/m^2, but the ratio converges rapidly to around 2 when the q_{u28} is larger than about 400 kN/m^2.

Other examples of the relationship between the q_{u7}, q_{u28} and q_{u91} on laboratory cement-stabilized soils are tabulated in Table 2.7(a) (Cement Deep Mixing Method Association, 1999) and those on field cement-stabilized soil by the pneumatic flow mixing method are tabulated in Table 2.7(b) (Ministry of Transport, The Fifth District Port Construction Bureau, 1999).

3.4.3 Curing temperature

Figure 2.29 shows the strength ratio of cement-stabilized soils, in which the Nagoya Port clay (w_L of 74.4%, w_P of 33.0% and w_i of 70.0%) was stabilized with ordinary Portland cement with cement factor, α, of 50 kg/m^3 (Ministry of Transport, The Fifth District Port Construction Bureau, 1999), where the stabilized soils were cured either in the air or underwater. In the figure, the strength of stabilized soil cured at an arbitrary temperature is normalized by the strength of the stabilized soil cured in the air at 20°C. The strength ratio shows a close relationship to the curing period. The strength of the stabilized soil cured underwater is about 20 to 30% larger than that in the air. The figure also shows that larger strength can be achieved at higher curing temperature, 30°C, even soon after mixing. However, small strength can be achieved at low curing temperature.

Table 2.7 Effect of curing period on unconfined compressive strength.

(a) on laboratory cement-stabilized soil (Cement Deep Mixing Method Association, 1999).

Ordinary Portland cement	Blast furnace slag cement type B

$q_u < 1,000 \text{ kN/m}^2$

| $q_{u7} - q_{u28}$ | $q_{u28} = 1.6q_{u7}$ | |
| $q_{u28} - q_{u91}$ | $q_{u91} = 1.1q_{u28}$ | |

$q_u < 3,000 \text{ kN/m}^2$

$q_{u7} - q_{u28}$	$q_{u28} = 1.49q_{u7}$	$q_{u28} = 1.56q_{u7}$
$q_{u7} - q_{u91}$	$q_{u91} = 1.97q_{u7}$	$q_{u91} = 1.95q_{u7}$
$q_{u28} - q_{u91}$	$q_{u91} = 1.44q_{u28}$	$q_{u91} = 1.20q_{u28}$

(b) on field cement-stabilized soil by pneumatic flow mixing method (Ministry of Transport, The Fifth District Port Construction Bureau, 1999).

Blast furnace slag cement type B

$q_{u28} - q_{u1h}$	$q_{u28} = 114.0q_{u1h}$
$q_{u28} - q_{u3h}$	$q_{u28} = 36.2q_{u3h}$
$q_{u28} - q_{u1}$	$q_{u28} = 4.5q_{u1}$
$q_{u28} - q_{u3}$	$q_{u28} = 2.1q_{u3}$
$q_{u28} - q_{u7}$	$q_{u28} = 1.6q_{u7}$
$q_{u91} - q_{u28}$	$q_{u91} = 1.1q_{u28}$

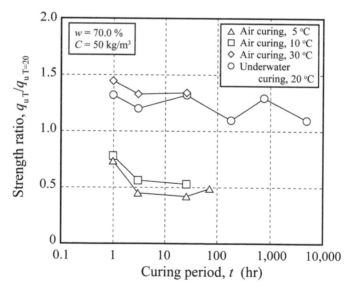

Figure 2.29 The effects of curing temperatures and curing conditions on the strength of cement-stabilized soil (Ministry of Transport, The Fifth District Port Construction Bureau, 1999).

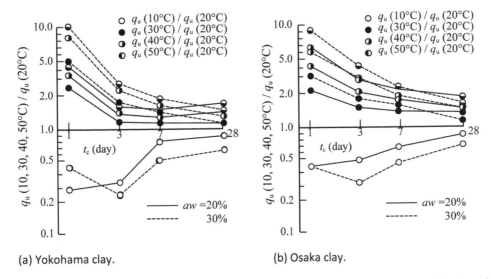

Figure 2.30 The effects of curing temperature on the strength of cement-stabilized soils (Saitoh et al., 1980).

The influence of curing temperature on the stabilized soil strength is shown in Figure 2.30, in which the two cement-stabilized soils, the Yokohama clay (w_L of 95.4% and w_P of 42.4%) and the Osaka clay (w_L of 79.4% and w_P of 40.2%) were cured at various temperatures for up to four weeks (Saitoh et al., 1980). In the figure, the strength of stabilized soil cured at an arbitrary temperature is normalized against the strength of stabilized soil cured at 20°C. The figure shows that larger strength can be achieved at a higher curing temperature. This influence of curing temperature is more dominant on short-term strength but it becomes less dominant as the curing period becomes longer.

The influence of curing temperature on the unconfined compressive strength of stabilized peat and silt was also investigated by Kido et al. (2009) and by Enami et al. (1985).

3.4.4 Maturity

In concrete engineering, the influences of curing temperature and curing period on the strength are often explained by the maturity index. Maturity is a concept that combines the effects of the curing period and temperature. Equation 2.3 shows four definitions of maturity proposed by the previous studies (M_1: general definition for cement-concrete, M_2: Nakamura et al., 2003, M_3: Åhnberg and Holm, 1984, and M_4: Babasaki et al., 1996a, 1996b). The correlation between the strength of stabilized soil and the logarithm of maturity, expressed differently, reveals that the curing temperature as an environmental condition does not have any significant effect on long term strength but has a considerable effect on short term strength.

$$M_1 = \sum (T_c - T_{c0}) \cdot t_c \tag{2.3a}$$

$$M_2 = 2.1^{(\theta - \theta_0)/10} \cdot t_c \tag{2.3b}$$

$$M_3 = \{20 + 0.5 \cdot (T_c - 20)\}^2 \cdot \sqrt{t_c} \qquad\qquad (2.3c)$$

$$M_4 = 2 \cdot exp\left(\frac{T_c - T_{c0}}{10}\right) \cdot t_c \qquad\qquad (2.3d)$$

where
M: maturity
T_c: curing temperature (°C)
T_{c0}: reference temperature (−10°C)
t_c: curing period (day)
θ: temperature (°C)
θ_0: reference temperature (°C)

Figure 2.31(a) shows the relationship between the unconfined compressive strength, q_u, and the curing temperature (Kitazume & Nishimura, 2009). The q_u value increases with the curing temperature, but the increasing ratio is dependent on the curing period. Figure 2.31(b) shows the relationship between the q_u and M_4 on the stabilized soil with various amount of cement. A unique relationship between the q_u and M_4 can be seen irrespective of the curing temperature and curing period, which can be formulated as Equation 2.4.

$$q_u = A \cdot \log M_4 + B \qquad\qquad (2.4)$$

where
M_4: maturity
A: parameter
B: parameter

According to a similar relationship between the q_u and M_4 on various types of soil: silt; peat (w_n of 456.9%); fine sand; loam (w_n of 109.9%); and clay (Enami et al., 1985, Horiuchi et al., 1984, Babasaki et al., 1984), the magnitude of the parameters A and B in Equation 2.4 are depended on the soil type (Babasaki et al., 1996a, 1996b).

3.4.5   Drying and wetting cycle

Figure 2.32 shows the unconfined compressive strengths of cement-stabilized soils subjected to the drying and wetting cycle (Kamon et al., 2005). In the tests, two volcanic cohesive soils (VCS-H and VCS-N) and the Fujinomori clay (Clay-F) were stabilized with ordinary Portland cement of 200 kg/m³. Their properties and initial water content are tabulated in Table 2.8. After 28 days' curing under the constant temperature of 20°C and relative humidity of 100%, the stabilized soils were subjected to the drying and wetting process up to 15 cycles, in which the soils were placed in a closed oven at 60 ± 3°C for 24 hours in the drying process, placed in room temperature for 1 hour, and immersed in distilled water at 20 ± 3°C for 23 hours in the wetting process. In the case of the Clay-F, the strength increases up to 6 cycles but rapidly decreases to a smaller strength than the initial strength with further cycles. In the cases of the

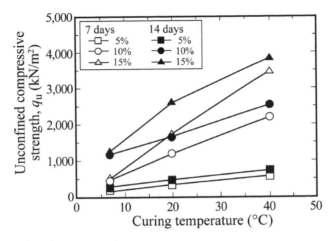

(a) Influence of curing temperature on q_u.

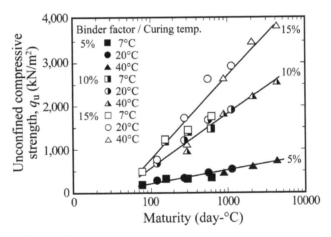

(b) Influence of maturity on q_u.

Figure 2.31 The influences of curing temperature and maturity on q_u (Kitazume & Nishimura, 2009).

volcanic soils, the strength is almost constant in the VCS-F soil but increases slightly with the cycles in the VCS-H soil.

3.4.6 Overburden pressure

Stabilized soil placed in the field is often subjected to overburden pressure during its curing period. Figure 2.33 shows the effect of overburden pressure during the curing period on the strength of the cement-stabilized soil, where the Ube clay (w_L of 45.4% and w_P of 20.1%) was stabilized with either ordinary Portland cement or cement-based special binder (SiO_2 of 15 to 20%, Al_2O_4 of more than 4.5%, CaO of 40 to 70%, and SO_4 of more than 4.0%) (Yamamoto et al., 2002). Figure 2.33 shows the relationship

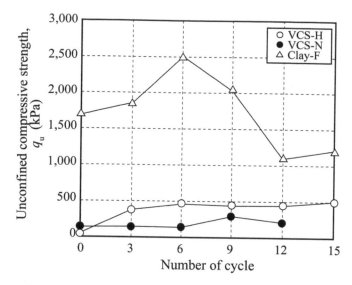

Figure 2.32 Unconfined compressive strength resulting from a number of drying and wetting cycles (Kamon et al., 2005).

Table 2.8 Physical and chemical properties of two volcanic cohesive soils, VCS-H and VCS-N, and a clay (Kamon et al., 2005).

	VCS-H	VCS-N	Clay-F
Particle density (g/cm³)	2.67	2.65	2.67
water content, w (%)	96.8	86.0	49.0
liquid limit, w_L (%)	122.1	114.8	51.2
plastic limit, w_P (%)	83.2	82.2	26.3
Grain size distribution			
sand fraction (2 mm–75 μm) (%)	52	64	49
silt fraction (75 μm–5 μm) (%)	23	21	23
clay fraction (under 5 μm-) (%)	25	15	28
pH	7.3	6.3	3.4
Cr(VI) leaching concentration	N.D.	N.D.	N.D.

between the unconfined compressive strength at 7 days' curing with the overburden pressure, σ'_v. The figure clearly shows that the strength increases almost linearly with the overburden pressure, irrespective of the type and amount of binder.

3.4.7   Soil disturbance/compaction

In some applications of the pneumatic flow mixing method, stabilized soil is temporarily placed at a provisional site, and is subsequently excavated and reclaimed at the final site after certain period for the beneficial use of the soil. Several research projects have been carried out to investigate the effect of disturbance on strength (e.g. Ministry of

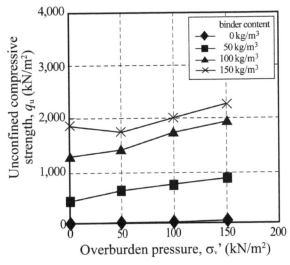

(a) Ordinary Portland cement.

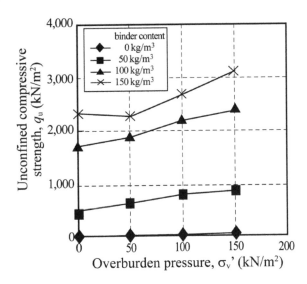

(b) Cement-based special binder.

Figure 2.33 The relationship between unconfined compressive strength, q_u, and overburden pressure, σ'_v (Yamamoto et al., 2002).

Transport, The Fifth District Port Construction Bureau, 1999; Makino et al., 2014, 2015). Figure 2.34 shows an example of the relationship between the unconfined compressive strength, q_u, of non-disturbed and disturbed stabilized soil and the curing period. In the tests, the kaolin clay (w_L of 77.5%, w_P of 30.3% and w_i of 120%) was stabilized with ordinary Portland cement with aw of 5 and 10%. After the mixing,

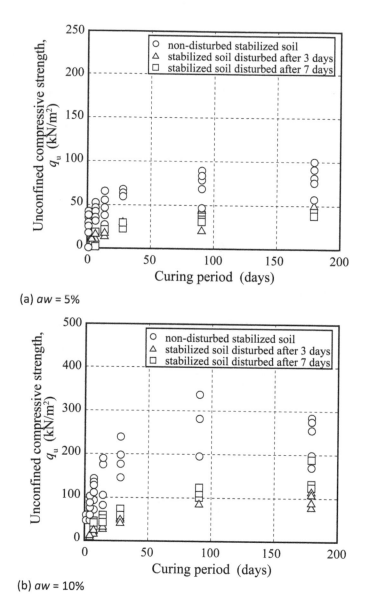

(a) $aw = 5\%$

(b) $aw = 10\%$

Figure 2.34 The effect of disturbance on the unconfined compressive strength of cement-stabilized soil (Makino et al., 2015).

the stabilized soil was molded by the tapping technique according to the Japanese Geotechnical Society Standard (Japanese Geotechnical Society Standard, 2009). In the case of the disturbed soil specimen, the stabilized soil mixture was stored and cured in an airtight plastic bag first, to avoid any change in water content. After 3 and 7 days' curing, the soil mixture in the bag was through disturbed and molded by the tapping

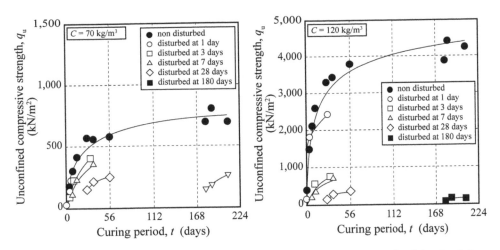

Figure 2.35 The effect of disturbance on the unconfined compressive strength of stabilized soil (Japan Cement Association, 2011).

technique in the same way as the un-disturbed soils. The unconfined compressive strength of the stabilized soils decreases considerably to 10 to 20% of that of the undisturbed stabilized soils, due to the disturbance, and it gradually increases to approximately 25 to 40% of that of the undisturbed stabilized soils with the curing period.

Figure 2.35 shows a similar test result, but in this case the stabilized soil is disturbed at several curing periods up to 180 days (Japan Cement Association, 2011). The figure shows that the strength increases to some extent as long as the soil is disturbed at an early stage in the curing, but its increase in strength is quite small when it is disturbed late in the curing period. The strength increases and reaches almost constant value at about 4 weeks in the case the cement content is relatively small. The strength gain is small when it is disturbed after the competition of cement hydration action.

Figure 2.36(a) shows the effect of the soil disturbance and compaction on the unconfined compressive strength, q_u, of cement-stabilized soil (Hino et al., 2007). In the test, the Hiroe Port clay (w_L of 144.4% and w_P of 52.0%) with an initial water content, w_i, of $1.5 \times w_L$ was stabilized with either blast furnace slag cement type B of 70 kg/m^3 or quicklime of 30 kg/m^3. A part of the stabilized soil was disturbed at 7 days' curing and compacted according to the Japanese standard (Japanese Geotechnical Society, 2009) with various compaction energies. Figure 2.36(a) shows that the stabilized soil strength considerably decreases to about 1/10 to 1/13 by the disturbance but increases gradually with the curing period after the compaction.

Figure 2.36(b) shows the relationship between unconfined compressive strength and compaction energy. In the case of the stabilized soil with blast furnace slag cement type B, the maximum strength can be seen at the compaction energy of 0.5 kJ/m^3, while the strength decreases with the compaction energy after the compaction energy exceeds 0.5 kJ/m^3. In the case of the quicklime stabilization, the maximum strength can be seen at 0.25 kJ/m^3. These results show the effect of the phenomenon of over-compaction on the strength of the stabilized soil.

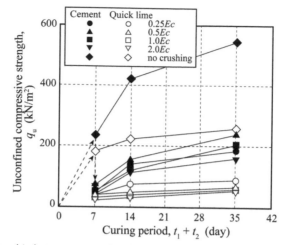

(a) Relationship between unconfined compressive strength and curing period.

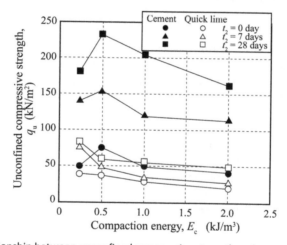

(b) Relationship between unconfined compressive strength and compaction energy.

Figure 2.36 The relationship between the unconfined compressive strength and applied compaction energy (Hino et al., 2007).

4 PREDICTION OF STRENGTH

In the admixture stabilization projects, including the pneumatic flow mixing project, the strength of field stabilized soil should be predicted and confirmed at various stages of planning, testing, design, and implementation. There are many proposed formulas to predict the laboratory strength and field strength of stabilized soil, which incorporate the various factors for the stabilization effect. Some of them are exemplified as follows

(Tsuchida & Tang, 2012, Horpibulsuk et al., 2000, Tang et al., 2000, Yanagihara et al., 2000, Miyazaki et al., 2003, Yoshida et al., 1977).

$$
\left.\begin{aligned}
&q_\mathrm{u} = aw + b \\
&q_\mathrm{u} = {a}/{(W/C)^x} + b \\
&q_\mathrm{u} = aC/w^x + b \\
&q_\mathrm{u} = \frac{10 \cdot Gs \cdot \rho_\mathrm{w} \cdot K\,(c - c_0)}{v^3} \\
&q_\mathrm{u} = \frac{A}{B^{(\overline{WC/C})}} \\
&q_\mathrm{u} = \frac{K\,(C - C_0)}{\left(w \cdot Gs/100 + 1\right)^2} \\
&q_\mathrm{u} = k_c \cdot \left(c^* - c_o^*\right) Y^3
\end{aligned}\right\}
\tag{2.5}
$$

where
a: parameter
A: parameter
b: parameter
B: parameter
c: cement factor (kg/m^3)
c^* : weight ratio, $W_c/(W_s + W_c)$
C : cement factor (kg/m^3)
c_0 : reference cement factor (kg/m^3)
c_0^* : reference weight ratio, $W_c/(W_s + W_c)$
C_0 : reference cement factor (kg/m^3)
Gs : specific gravity of soil particle
K: parameter
k_c: parameter
q_u: unconfined compressive strength (kN/m^2)
Vc : volume of cement (m^3)
Vs : volume of soil (m^3)
Vv : volume of void and water (m^3)
w: water content (%)
Wc : dry weight of cement (kg)
W/C: water to cement ratio
WC/C: clay water content/cement ratio
Ws : dry weight of soil (kg)
x: parameter
Y: volume ratio, $(Vs+Vc)/(Vs+Vc+Vv)$
ρ_w: density of water (g/cm^3)

For long term strength, the equation 2.6 is given by Yanagihara et al. (2000):

$$
\left.\begin{aligned}
&q_\mathrm{u} = a \times t^b \\
&q_\mathrm{u} = a \times \log(t) + b
\end{aligned}\right\}
\tag{2.6}
$$

where
a: parameter
b: parameter
t: curing period

As shown in Equations 2.5 and 2.6, the equations consist of several parameters which may be influenced by many factors tabulated in Table 2.1. The strengths of laboratory and field stabilized soil can be predicted by the equations if the appropriate magnitude of these parameters can be estimated precisely. However, we are not yet at the stage where we can estimate these parameters well enough to predict the laboratory strength with a reasonable level of accuracy. There is no widely-applicable formula for estimating the strength which incorporates all the relevant factors.

Because of this, the most precise prediction can be now made by performing a laboratory mix test and then estimating the field strength on the basis of laboratory mix test results and past experience. In large scale projects, laboratory mix test results are often confirmed by field trial execution of stabilized soil at the construction site. For small scale works, reference is made to previous soil stabilization works done in similar soil conditions.

Nevertheless, the information compiled in the present chapter is extremely valuable in planning pneumatic flow mixing work and also for interpreting laboratory mix test results if properly used by an experienced engineer.

REFERENCES

Åhnberg, H. & Holm, G. (1984) On the influence of curing temperature on the strength of lime and cement stabilised soils. *Swedish Geotechnical Institute Report*. Vol. 30. pp. 93–146 (in Swedish).

Babasaki, R., Terashi, M., Suzuki, K., Maekawa, J., Kawamura, M. & Fukazawa, E. (1996a) Factors influencing the strength of improved soils. *Proc. of the Symposium on Cement Treated Soils*. pp. 20–41 (in Japanese).

Babasaki, R., Terashi, M., Suzuki, T., Maekawa, A., Kawamura, M. & Fukazawa, E. (1996b) Japanese Geotechnical Society Technical Committee Reports: Factors influencing the strength of improved soil. *Proc. of the 2nd International Conference on Ground Improvement Geosystems*. Vol. 2, pp. 913–918.

Babasaki, R., Kawasaki, T., Niina, A., Munechika, Y. & Saitoh, S. (1980) Study of the deep mixing method using cement hardening agent (Part 9). *Proc. of the 15th Annual Conference of the Japanese Society of Soil Mechanics and Foundation Engineering*. pp. 713–716 (in Japanese).

Babasaki, R., Saito, S. & Suzuki, Y. (1984) Temperature characteristics of cement improved soil and temperature analysis of ground improved using the deep cement mixing method. *Proc. of the Symposium on Strength and Deformation of Composite Ground*. pp. 33–40 (in Japanese).

Cement Deep Mixing Method Association (1999) *Cement Deep Mixing Method (CDM) Design and Construction Manual*. (in Japanese).

Coastal Development Institute of Technology (2008a) *Technical Manual of Deep Mixing method for marine works, revised version*. 289p. (in Japanese).

Coastal Development Institute of Technology (2008b) *Technical Manual of Pneumatic Flow Mixing Method*, revised version. Daikousha Publishers, 188p. (in Japanese).

Enami, A., Yoshida, M., Hibino, S., Takahashi, M. & Akitani, K. (1985) In situ measurement of temperature in soil cement columns and influence of curing temperature on unconfined compressive strength of soil cement. *Proc. of the 20th Annual Conference of the Japanese Society of Soil Mechanics and Foundation Engineering.* pp. 1737–1740 (in Japanese).

Hayashi, H., Noto, S. & Toritani, N. (1989) Cement improvement of Hokkaido peat. Symposium on High Organic Soils. *Proc. of the Symposium on High Organic Soils.* pp. 101–106 (in Japanese).

Hino, Y., Suetsugu, D., Ikari, Y. & Ohga, T. (2007) Unconfined compressive strength, cone index and compaction properties of embankment materials by dredged-stabilized clay. *Proc. of the 42nd Annual Conference of the Japanese Geotechnical Society.* pp. 615–616 (in Japanese).

Horiuchi, N., Ito, M., Morita, T., Yoshihara, S., Hisano, T., Hanazono, H. & Tanaka, T. (1984) Strength of soil mixture under lower temperatures. *Proc. of the 19th Annual Conference of the Japanese Society of Soil Mechanics and Foundation Engineering.* pp. 1609–1610 (in Japanese).

Horpibulsuk, H., Miura, N. & Nagaraj, T.S. (2000) A new method for predicting strength of cement stabilized clays. *Prof of the International Conference of the Coastal Geotechnical Engineering in Practice.* Vol. 1, pp. 605–610.

Ingles, O.G. & Metcalf, J.B. (1972) *Soil stabilization, Principles and Practice.* Butterworth.

Japan Cement Association (2009) Committee report on soil stabilization of dredged soil. *Internal Report of Japan Cement Association.* 57p. (in Japanese).

Japan Cement Association (2011) Influence of disagglomeration on strength of dredged soil stabilized with soil cement. *Cement & Concrete.* No. 769, pp. 2–7 (in Japanese).

Japan Cement Association (2012) *Soil Improvement Manual using Cement Stabilizer (4th edition).* Japan Cement Association. 442p. (in Japanese).

Japan Lime Association (2009) *Technical Manual on Ground Improvement using Lime.* Japan Lime Association. 176p. (in Japanese).

Japanese Geotechnical Society (2000) *Practice for making and curing stabilized soil specimens without compaction, JGS T 0821-2000.* Japanese Geotechnical Society. pp. 308–316 (in Japanese).

Japanese Geotechnical Society (2009) *Practice for making and curing stabilized soil specimens without compaction. JGS 0821-2009.* Japanese Geotechnical Society. Vol. 1, pp. 426–434 (in Japanese).

Japanese Industrial Standard (2006) *Portland blast-furnace slag cement, JIS R 5211: 2006.* (in Japanese).

Japanese Industrial Standard (2009) *Portland cement, JIS R 5210: 2009.* (in Japanese).

Japanese Society of Soil Mechanics & Foundation Engineering (1990) *Practice for making and curing non-compacted stabilized soil specimens. JGS T 821-1990.* Japanese Society of Soil Mechanics and Foundation Engineering. (in Japanese).

Kamon, M., Inui, T. & Shoji, Y. (2005) Experimental study on the long-term environmental impact caused by the cement stabilization/solidification of soft ground. *Annual of Disaster Prevention Research Institute,* Kyoto University. No.48 B, (in Japanese).

Kawasaki, T., Niina, A., Saitoh, S. & Babasaki, R. (1978) Studies on engineering characteristics of cement-base stabilized soil. *Takenaka Technical Research Report.* Vol. 19, pp. 144–165 (in Japanese).

Kawasaki, T., Niina, A., Saitoh, S., Suzuki, Y. & Honjyo, Y. (1981) Deep mixing method using cement hardening agent. *Proc. of the 10th International Conference on Soil Mechanics and Foundation Engineering.* Vol. 3, pp. 721–724.

Kido, Y., Hishimoto, S., Hayashi, H. & Hashimoto, H. (2009) Effects of curing temperatures on the strength of cement-treated peat. *Proc. of the International Symposium on Deep Mixing and Admixture Stabilization.* pp. 151–154.

Kitazume, M. & Nishimura, S. (2009) Influence of specimen preparation and curing conditions on unconfined compression behaviour of cement-treated clay. *Proc. of the International Symposium on Deep Mixing and Admixture Stabilization.* pp. 155–160.

Kitazume, M. & Satoh, T. (2003) Development of pneumatic flow mixing method and its application to Central Japan International Airport construction. *Journal of Ground Improvement.* Vol. 7, No. 3, pp. 139–148.

Kiyota, M., Tutumi, T., Ohta, K. & Hirayama, Y. (2003) About the characteristic of the improvement soil with the slow-setting soil stabilizer. *Proc. of the 38th Annual Conference of the Japanese Geotechnical Society.* pp. 799–800 (in Japanese).

Makino, M., Takeyama, T. & Kitazume, M. (2014) Laboratory tests on the influence of soil disturbance on the material properties of cement-treated soil. *Proc. of the 9th International Symposium on Lowland Technology.* pp. 150–155.

Makino, M., Takeyama, T. & Kitazume, M. (2015) The influence of soil disturbance on material properties and micro-structure of cement-treated soil. *International Journal of Lowland Technology.*

Miki, H., Kudara, K. & Okada, Y. (1984) Effect of humic acid on soil stabilization (Part 2). *Proc. of the 39th Annual Conference of the Japan Society of Civil Engineers.* pp. 407–408 (in Japanese).

Ministry of Transport, The Fifth District Port Construction Bureau (1999) *Pneumatic Flow Mixing Method.* Yasuki Publishers. 157p. (in Japanese).

Miyazaki, Y., Tang, Y.X., Ochiai, H., Yasufuku, N., Omine, K. & Tsuchida, T. (2003). Utilization of cement treated dredged soil with working ship. *Journal of the Japan Society of Civil Engineers.* No.750/III-65, pp. 193–204, 2003 (in Japanese).

Nakamura, T., Saitoh, S. & Babasaki, R. (2003) Early stage estimation for 28 day unconfined compressive strength of cement-improved soils using an accelerated curing test. *Proc. of the 38th Annual Conference of the Japanese Geotechnical Society.* (in Japanese).

Nakamura, M., Akutsu, H. & Sudo, F. (1980) Study of improved strength based on the deep mixing method (Report 1). *Proc. of the 15th Annual Conference of the Japanese Society of Soil Mechanics and Foundation Engineering.* pp. 1773–1776 (in Japanese).

Nakamura. M., Matsuzawa, S. & Matsushita, M. (1982) Studies on mixing efficiency of stirring wings for deep mixing method. *Proc. of the 17th Annual Conference of the Japanese Society of Soil Mechanics and Foundation Engineering.* Vol. 2, pp. 2585–2588 (in Japanese).

Niina, A., Saitoh, S., Babasaki, R., Miyata, T. & Tanaka, K. (1981) Engineering properties of improved soil obtained by stabilizing alluvial clay from various regions with cement slurry. *Takenaka Technical Research Report.* Vol. 25, pp. 1–21 (in Japanese).

Niina, A., Saitoh, S., Babasaki, R., Tsutsumi, I. & Kawasaki, T. (1977) Study on DMM using cement hardening agent (Part 1). *Proc. of the 12th Annual Conference of the Japanese Society of Soil Mechanics and Foundation Engineering.* pp. 1325–1328 (in Japanese).

Okada, Y., Kudara, K. & Miki, H. (1983) Effect of humic acid on soil stabilization. *Proc. of the 38th Annual Conference of the Japan Society of Civil Engineering.* pp. 467–468 (in Japanese).

Okano, Y., Yoda, H., Sasayama, T., Nakano, M., Yamada, S., Nonoyama, S. & Sakai, T. (2012) Mechanical properties of cement stabilized soil with different physical properties. *Proc. of the Annual Research Meeting, Chubu Branch, Japan Society of Civil Engineering.* pp. 169–170 (in Japanese).

Okumura, T., Terashi, M., Mitsumoto, T., Yoshida, T. & Watanabe, M. (1974) Deep-Lime-Mixing method for soil stabilization (3rd Report). *Report of the Port and Harbour Research Institute.* Vol. 13, No. 2, pp. 3–44 (in Japanese).

Saitoh, S. (1988) Experimental study of engineering properties of cement improved ground by the deep mixing method. *Doctoral thesis, Nihon University.* 317p. (in Japanese).

Saitoh, S., Niina, A. & Babasaki, R. (1980) Effect of curing temperature on the strength of treated soils and consideration on measurement of elastic modules. *Proc. of the Symposium on Testing of treated Soils, Japanese Society of Soil Mechanics and Foundation Engineering.* pp. 61–66 (in Japanese).

Saitoh, S., Suzuki, Y. & Shirai, K. (1985) Hardening of soil improved by the deep mixing method. *Proc. of the 11th International Conference on Soil Mechanics and Foundation Engineering.* Vol. 3, pp. 1745–1748.

Sasayama, T., Yoda, H., Horiuchi, S., Nakano, M. & Yamada, S. (2011) Mechanical properties of cement stabilized soil with different particle size distribution. *Proc. of the Annual Research Meeting, Chubu Branch, Japan Society of Civil Engineering.* pp. 193–194 (in Japanese).

Tang, Y.X., Miyazaki, Y. & Tsuchida, T. (2000) Advanced reuses of dredging by cement treatment in practical engineering. *Prof of the International Conference of the Coastal Geotechnical Engineering in Practice.* Vol. 1, pp. 725–731.

Terashi, M. (1997) Theame Lecture: Deep Mixing Method - Brief State of the Art. *Proc. of the 14th International Conference on Soil Mechanics and Foundation Engineering.* Vol. 4, pp. 2475–2478.

Terashi, M., Okumura, T. & Mitsumoto, T. (1977) Fundamental properties of lime-treated soils. *Report of the Port and Harbour Research Institute.* Vol. 16, No. 1, pp. 3–28 (in Japanese).

Terashi, M., Tanaka, H. & Okumura, T. (1979) Engineering properties of lime-treated marine soils and D.M.M. method. *Proc. of the 6th Asian Regional Conference on Soil Mechanics and Foundation Engineering.* Vol. 1, pp. 191–194.

Terashi, M., Tanaka, H., Mitsumoto, T., Honma, S. & Ohhashi, T. (1983) Fundamental properties of lime and cement treated soils (3rd Report). *Report of the Port and Harbour Research Institute.* Vol. 22, No. 1, pp. 69–96 (in Japanese).

Terashi, M., Tanaka, H., Mitsumoto, T., Niidome, Y. & Honma, S. (1980) Fundamental properties of lime and cement treated soils (2nd Report). *Report of the Port and Harbour Research Institute.* Vol. 19, No. 1, pp. 33–62 (in Japanese).

Thompson, R. (1966) Lime reactivity of Illinois soils. *Proc. of the American Society of Civil Engineering.* 92 (SM-5).

Tsuchida, T. & Tang, Y.X. (2012) A consideration on estimation of strength of cement-treated marine clays. *Japanese Geotechnical Journal.* Vol. 7, No. 3, pp. 435–447 (in Japanese).

Tsuchida, T., Tang, Y.X., Shimakawa, N. & Abe, T. (2013) Study on the strength–time relationship of cement-treated marine clays with high water contents. *Japanese Geotechnical Journal.* Vol. 8, No. 1, pp. 53–70 (in Japanese).

Udaka, K., Tsuchida, T., Imai, Y. & Tang, Y.X. (2013) Compressive characteristics of reconstituted marine clays with developed structures by adding a small amount of cement. *Japanese Geotechnical Journal.* Vol. 8, No. 3, pp. 425–439 (in Japanese).

Watabe, Y., Tsuchida, T., Hikiyashiki, H. & Furuno, T. (2001) Mechanical and material properties of dredged soil treated with poor quality of cement. *Report of the Port and Harbour Research Institute.* Vol. 40, No. 2, pp. 3–21 (in Japanese).

Yamamoto, T., Suzuki, M., Okabayashi, S., Fujino, H., Taguchi, T. & Fujimoto, T. (2002) Unconfined compressive strength of cement-stabilized soil cured under an overburden pressure. *Journal of Geotechnical Engineering, Japan Society of Civil Engineers.* No. 701/3-58, pp. 387–399 (in Japanese).

Yanagihara, M., Horiuchi, S. & Kawaguchi, M. (2000) Long-term stability of coal-fly-ash slurry man-made island. *Prof of the International Conference of the Coastal Geotechnical Engineering in Practice.* Vol. 1, pp. 763–769.

Yoshida, T., Kubo, H. and Sumida, K (1977). Investigation on treatment of sludge (Part 3) Application of pF water to relationship between strength of cement treated soil and water cement ratio. *Proc. of the 12th Annual Conference of the Japanese Society of Soil Mechanics and Foundation Engineering.* pp. 1309–1312 (in Japanese).

Chapter 3

Engineering properties of stabilized soils

I INTRODUCTION

The engineering properties of lime- and cement-stabilized soils have been extensively studied by highway engineers since the 1960s. However, the purpose of their stabilization was to improve sub-base or subgrade materials and the stabilization was characterized by the low water content of the original soil and the small amount of binder. Mixing a few per cent of binder with respect to the dry weight of soil is enough to change the physical properties of soil in order to enable efficient compaction following mixing.

As shown in Table 1.3, soils appropriate for the pneumatic flow mixing method are very soft dredged clay, organic soil, and soft alluvial soil, which usually have a water content exceeding their liquid limit. The amount of binder needed is usually relatively small to achieve a target strength of the order of 100 to 200 kN/m^2. Compaction of a nearly-saturated soil-binder mixture is ineffective and practically impossible to carry out soon after mixing, whereas it can be conducted in some cases after certain curing period for beneficial use, as explained in Chapter 2. The purpose of stabilization in the pneumatic flow mixing method is usually to manufacture a large strength soil material for land reclamation, backfilling, shallow stabilization, *etc*. Due to these differences in manufacturing process, and in the expected functions of stabilized soil, the fundamental engineering properties of lime- and cement-stabilized clays and sands have been studied in detail in Japan.

Although the magnitude of strength gain by stabilization is influenced by various factors, including the type of binder (Chapter 2), the engineering properties of lime- and cement-stabilized soils are quite similar. Since cement and a cement-based type special binder have both been used in the pneumatic flow mixing method, properties of cement-stabilized soil will be mainly described in this chapter.

The descriptions in this chapter are mostly based on the research done in Japan or on accumulated experience on Japanese soils, machines and mixing systems. The soil properties introduced here may not be directly applicable in the other parts of the world.

Chapter 3 describes the engineering properties of soil stabilized mainly with cement. Transport and placement ability are key factors for the execution of the method, which is influenced by the engineering properties of the freshly stabilized soil, soon after the mixing. This chapter introduces not only the properties of hardened stabilized soil but also the properties of fresh stabilized soil. There are quite a lot of test data on the properties of stabilized soil for the deep mixing method (Kitazume & Terashi, 2013). However, the target strength of the pneumatic flow mixing method is smaller than that of the deep mixing method, while the water content is high and the amount of cement is small. Therefore the properties of the soft stabilized soil are somewhat different from the hard stabilized soil. Correlation between the unconfined compressive strength of stabilized soil and its other engineering characteristics will be of benefit in the understanding of stabilized soil, regardless of the execution system and type of application. The characteristics of *in-situ* stabilized soil, such as the relationship between the average field strength and the laboratory strength, and the variability of field strengths, are important information for geotechnical design. This is discussed, but based only upon the experience gained by the execution system used in Japan. This is because the quality of *in-situ* stabilized soil heavily depends upon the mixing machine and process. The chapter concludes that it is necessary for contractors to take responsibility for accumulating information on the quality of *in-situ* stabilized soil produced by their proprietary system.

2 PROPERTIES OF STABILIZED SOIL MIXTURE BEFORE HARDENING

2.1 Physical properties

2.1.1 Change in consistency of the soil–binder mixture before hardening

The water content of stabilized soil decreases in many cases due to the hydration of quicklime and cement. At the same time, the consistency of the soil–binder mixture changes from that of the original soil due to the ion exchange. As no research result was found on cement-stabilized soil, Figure 3.1 shows an example of the effect of quicklime stabilization on the consistency of the soil–binder mixture measured at three hours

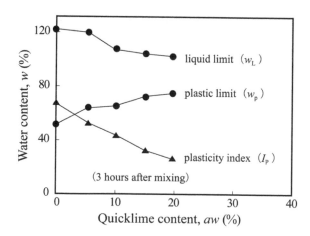

Figure 3.1 Change of consistency in soil resulting from quicklime stabilization (Japan Lime Association, 2009).

after mixing (Japan Lime Association, 2009). The liquid limit, w_L, decreases with the quicklime content, while the plastic limit, w_P, increases. As a result, the plasticity index, I_p, sharply decreases with increasing quicklime content.

2.2 Mechanical properties (strength characteristics)

2.2.1 Change in flow value

The flow value of the stabilized soil mixture before hardening is influenced by the mixing conditions, and it is usually quite a small value compared with that of the original soil, even with the same water content. Figure 3.2(a) shows the relationship between the water content of the original soil and the stabilized soil, and their flow values (Ministry of Transport, The Fifth District Port Construction Bureau, 1999). The Nagoya Port soil (w_L of 74.4% and w_P of 33.0%) was stabilized with ordinary Portland cement with a cement factor of 30, 50 and 70 kg/m^3, in which the W/C ratio of cement slurry is 100%. The flow value of the stabilized soil mixture was measured at 10 minutes after mixing by the Japanese Industrial Standard (JIS A 313). Figure 3.2(a) reveals that the flow value increases with the water content of stabilized soil mixture, irrespective of the cement factor. In the figure, the flow value of the original soil is plotted together; this also increases with the water content and is somewhat larger than those of the stabilized soil mixture.

Figure 3.2(b) shows the relationship between the flow value and the shear strength of the cement-stabilized soil mixture, which were measured at 10 minutes after mixing. It shows that the shear strength exponentially decreases with the flow value and a unique relationship between them can be seen irrespective of the cement factor (Ministry of Transport, The Fifth District Port Construction Bureau, 1999).

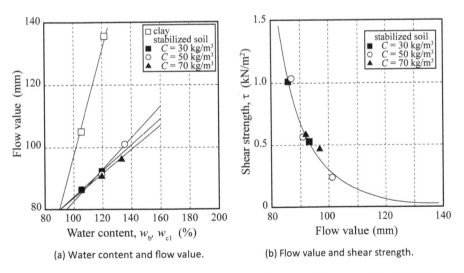

(a) Water content and flow value.   (b) Flow value and shear strength.

Figure 3.2 Characteristics of fresh cement-stabilized soil (Ministry of Transport, The Fifth District Port Construction Bureau, 1999).

2.2.2   Change in shear strength

The initial strength gain of stabilized soil mixture is one of the key properties to influence the slope angle of stabilized soil mixture after placement, which will be described in Section 5 of this chapter. Figure 3.3(a) shows the shear stress gain, τ_v (Watabe et al., 2001). The Tachibana Bay clay (w_L of 40.8% and w_P of 20.8%) was prepared to have an initial water content of $1.6 \times w_L$ and then stabilized with ordinary Portland cement with a cement factor of $50 \, kg/m^3$. Figure 3.3(b) shows the strength increase of the cement-stabilized soil soon after mixing (Watabe et al., 2001). In the tests, the Tachibana Bay clay (w_L of 40.8% and w_P of 20.8%) was prepared to have an initial water content of $1.6 \times w_L$ (65.3%) and stabilized with ordinary Portland cement with a cement factor, α, of $50 \, kg/m^3$. Soon after the mixing, the shear strength, τ_v, was measured with a hand vane shear apparatus at several elapsed times up to 200 minutes. The shear strength of the original soil with a water content of $1.6 \times w_L$ was also measured and plotted in the figure. The shear strength of the stabilized soil mixture just after mixing is larger than that of the original soil, which is probably due to reduced fluidity as a result of adding cement powder. The shear strength, τ_v, remains a small value up to about 30 minutes but increases rapidly after that, due to the progress of the cement hydration effect.

2.2.3   Stress – strain curve

Figure 3.4 shows the stress–strain curves on cement-stabilized Tachibana Bay clay (Watabe et al., 2001). The Tachibana Bay clay (w_L of 40.8% and w_P of 20.8%) was prepared to have an initial water content of $1.6 \times w_L$ and stabilized with ordinary Portland cement with a cement factor of $50 \, kg/m^3$. The soil samples were subjected to the unconfined compression test at various curing periods. In the figure, the stress–strain curve on the original soil was also plotted, which was consolidated with the pressure of 49 kPa. The compressive stress, q, increases gradually with the axial strain to reach a constant value of about 25 kPa, which shows the ductile behavior. In the case of the stabilized soils, on the other hand, the compressive stress increases rapidly with the axial strain to a peak strength, which is followed by small residual strength. The peak compressive stress increases quickly and the axial strain at failure decreases quickly with the curing period. The brittle behavior with a large peak strength, small strain at failure and a small residual strength becomes more dominant with the curing period.

Figure 3.4(b) shows the effect of confined pressure on the stress-strain curves on cement-stabilized clay (Watabe et al., 2001). The stabilized clay was allowed to isotropically consolidate under the confined pressure of 49 kPa and was then subjected to undrained compression. The maximum strength increases with the curing period as similar to those of the unconfined compression test (Figure 3.4(a)), but the compressive stress decreases slightly to a relatively large residual strength. The pore water pressure also increases with the axial strain, but its magnitude is almost of the same order irrespective of the curing period even the peak strength is influenced by the curing period very much.

2.3   Mechanical properties (consolidation characteristics)

Figure 3.5(a) shows e-$\log p$ curves of the laboratory cement-stabilized soils, in which the Tachibana Bay clay (w_L of 40.8% and w_P of 20.8%) was stabilized with ordinary

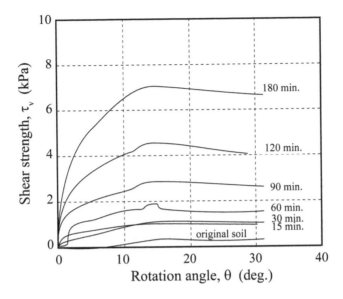

(a) Relationship between shear stress and rotation angle of the vane blade.

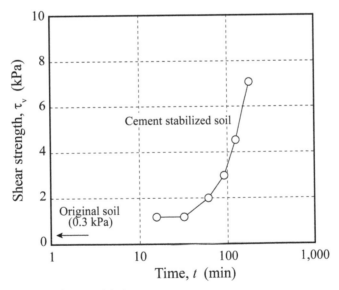

(b) Shear strength gain with time.

Figure 3.3 Strength increase soon after cement mixing (Watabe et al., 2001).

Portland cement with a cement factor, α, of $150 \, \text{kg/m}^3$ (Watabe et al., 2001). In the laboratory tests, the stabilized soil samples of 20 mm height and 60 mm diameter were consolidated one dimensionally by the constant strain rate consolidation test. The figure shows a sharp bend in the curves irrespective of the curing period.

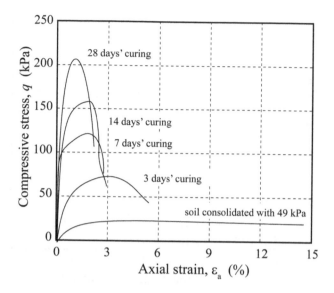

(a) Stress - strain curves of cement-stabilizedsoils in an unconfined compression test.

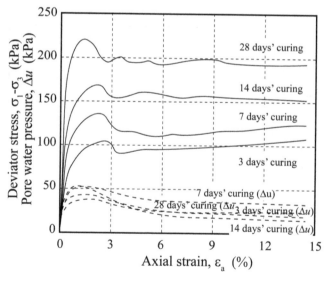

(b) Stress and pore water pressure strain curves of cement-stabilizedsoils in a consolidated undrained compression test.

Figure 3.4 Stress and pore water pressure strain curves of cement-stabilized clays (Watabe et al., 2001).

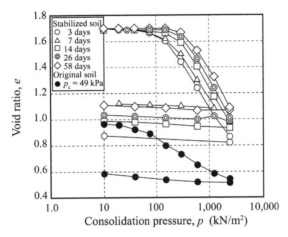

(a) e-log p curve.

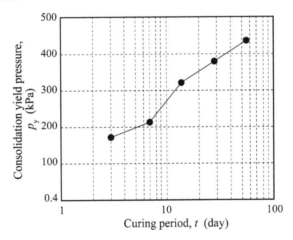

(b) Relationship between the consolidation yield pressure, p_y, and curing.

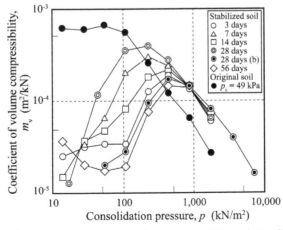

(c) Relationship between the coefficient of volume compressibility and consolidation pressure

Figure 3.5 Consolidation properties of cement-stabilized soil (Watabe et al., 2001).

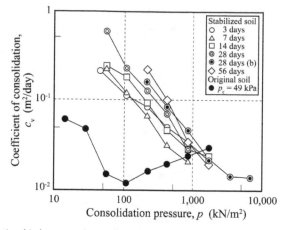

(d) Relationship between the coefficient of consolidation and consolidation pressure.

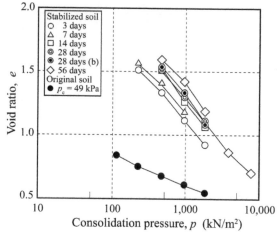

(e) Relationship between the void ratio and consolidation pressure.

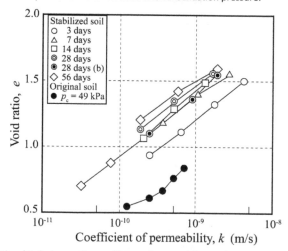

(f) Relationship between the coefficient of permeability and void ratio.

Figure 3.5 (Continued).

The consolidation yield pressure at the sharp bend is larger for the longer curing period, while the compression index, Cc, and the swelling index, Cs, are almost same irrespective of the curing period. Figure 3.5(b) shows the relationship between the consolidation yield pressure, p_y, and the curing period, where the p_y increases almost linearly with the logarithm of the curing period. Figures 3.5(c) and 3.5(d) show the relationships between the coefficient of volume compressibility, m_v, the coefficient of consolidation, c_v, and the consolidation pressure. The coefficient of volume compressibility, m_v, is around 2.0×10^{-5} m²/kN at the small consolidation pressure and increases with the consolidation pressure to the peak value at around the consolidation yield pressure, p_y. The consolidation pressure when the m_v increases becomes large with the curing period, while the m_v value at the large consolidation pressure is almost same irrespective of the curing period. The coefficient of consolidation, c_v, decreases with the consolidation pressure to a constant value. Figures 3.5(e) and 3.5(f) show the relationships between the void ratio, e, consolidation pressure, p, and coefficient of permeability, k, which are measured in the consolidation test. The permeability, k, decreases with decreasing void ratio. The k value of the stabilized soil is larger than that of the original soil, which is due to the large void ratio of the stabilized soil.

Figure 3.6 shows the coefficient of permeability of stabilized soils, which was measured by a laboratory falling head permeability test (Ministry of Transport, The Fifth District Port Construction Bureau, 1999). The Nagoya Port clay (w_L of 68.7% and w_P of 29.6%) was prepared to an initial water content of 96.2% and stabilized with ordinary Portland cement of 100 kg/m³ (Ministry of Transport, The Fifth District Port Construction Bureau, 1999). The figure shows that the coefficient of permeability, k, is very much smaller than that of the original soil, and of the order of about 10^{-7} cm/s. The permeability gradually decreases with the progress of cement hydration.

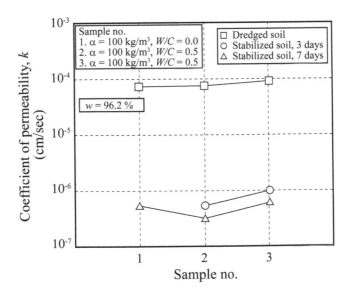

Figure 3.6 The relationship between permeability of stabilized soil and mixing condition (Ministry of Transport, The Fifth District Port Construction Bureau, 1999).

3 PROPERTIES OF STABILIZED SOIL AFTER HARDENING

3.1 Physical properties

3.1.1 Change in water content

The water content of cement-stabilized soil after cement hydration can be estimated by Equation (3.1). The water-to-soil ratio required for cement hydration, λ, is dependent upon the type and composition of the cement, but can be estimated to be about 25 to 40% of the dry weight of cement.

$$w_t = \frac{w_0 + (W/C - \lambda) \cdot aw/100}{100 + (100 - \lambda) \cdot aw/100} \times 100 \tag{3.1}$$

where
aw: cement content (%)
w_0: water content of original soil (%)
w_t: water content of stabilized soil (%)
W/C: water to cement ratio (%)
λ: water to soil ratio required for cement hydration (25 to 40%)

Figure 3.7 shows the water contents of laboratory cement-stabilized soils, in which the Shinagawa clay (w_L of 62.6%, w_P of 23.1% and w_i of 76.5%) was stabilized with

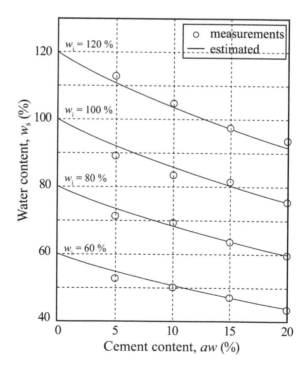

Figure 3.7 The change in water content brought about by cement stabilization (Kawasaki et al., 1978).

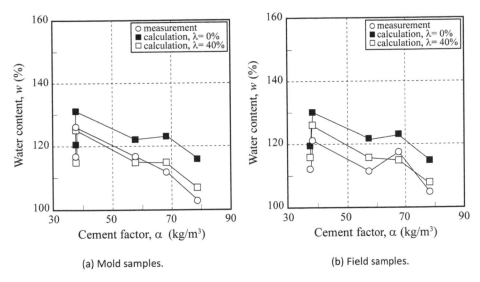

Figure 3.8 The change in water content brought about by cement stabilization (Ministry of Transport, The Fifth District Port Construction Bureau, 1999).

ordinary Portland cement with a cement content, aw, of 5, 10, 15 and 20% (Kawasaki et al., 1978). The water content of the stabilized soil decreases gradually with the cement content. In the figure, the estimated values by Equation (3.1) with λ of 25% are also plotted. The estimated values coincide with the measured values very well.

Figure 3.8 shows the water content change in a cement-stabilized soil at 28 days' curing, in which the Nagoya Port clay (w_L of 74.4%, w_P of 33.0% and w_i of 127.1%) was stabilized with ordinary Portland cement with various cement factors by one of the pneumatic flow mixing methods in the field (Ministry of Transport, The Fifth District Port Construction Bureau, 1999). The stabilized soil mixture was collected at the outlet of the pipeline and placed in a plastic mold and cured in a laboratory (mold samples), while the stabilized soil placed and cured in the field was core sampled at 28 days' curing (field samples). It was found that the water content of the stabilized soil decreases gradually with the cement content. In the figure, the calculated values by Equation (3.1) with λ of 0% (no hydration) and λ of 40% (full hydration is assumed) are also plotted. The measured values are about 5 to 10% smaller than the calculation for the no hydration condition, but are relatively well coincident with the calculations for the full hydration.

3.1.2   Change in density

The saturated density of cement-stabilized soil can be calculated by Equation 3.2.

$$\rho_t = \frac{100 + w_0 + (100 + W/C) \cdot aw/100}{\frac{100}{G_s} + \left(\frac{100}{G_c} + \frac{W/C}{G_w}\right) \cdot aw/100 + \frac{w_0}{G_w}} \times \rho_w \tag{3.2}$$

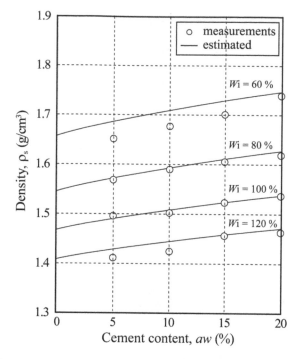

Figure 3.9 The change of density brought about by cement stabilization (Kawasaki et al., 1978).

where
aw: cement content (%)
G_c: specific gravity of binder
G_s: specific gravity of soil particle
G_w: specific gravity of water
w_o: water content of original soil (%)
W/C: water to cement ratio (%)
ρ_t: density of stabilized soil (g/cm^3)
ρ_w: density of water (g/cm^3)

Figure 3.9 shows the density of the laboratory cement-stabilized soils, in which the Kawasaki clay (w_L of 62.6% and w_P of 23.1%) was stabilized with ordinary Portland cement with various cement contents, aw of 5, 10, 15 and 20% (Kawasaki et al., 1978). The density of the stabilized soil increases gradually with the cement content. In the figure, the estimated values by Equation (3.2) are also plotted, which coincide with the measured values well.

Figure 3.10 also shows the unit weight of the field cement-stabilized soils, in which the Nagoya Port clay (w_L of 74.4%, w_P of 33.0% and w_i of 127.1%) was stabilized with ordinary Portland cement of various cement factors by one of the pneumatic flow mixing methods in the field (Ministry of Transport, The Fifth District Port Construction Bureau, 1999). The stabilized soil mixture was collected at the outlet of the pipeline and placed in a plastic mold and cured in a laboratory (mold samples), while

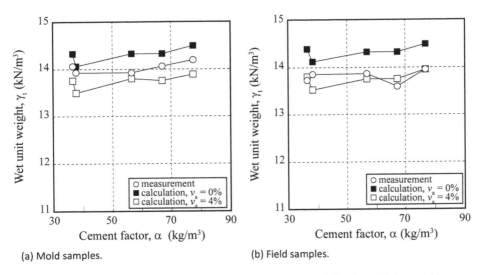

(a) Mold samples.   (b) Field samples.

Figure 3.10 The change of density brought about by cement stabilization (Ministry of Transport, The Fifth District Port Construction Bureau, 1999).

the stabilized soil placed and cured in the field was core sampled at 28 days' curing (field samples). The unit weight of the stabilized soils increases gradually with the cement factor. In the figure, the estimated values by Equation 3.2 with V_a of 0% (fully saturated condition) are also plotted, which overestimates the measured value. The calculation for taking account of the entrapped air of 4.7% of the total volume is also plotted, which was measured in a case history of the pneumatic flow mixing, which coincides with the measured value relatively well.

3.1.3 Change in consistency

Figure 3.11 shows the effect of the cement stabilization on the consistency of stabilized soil (Watabe & Tanaka, 2012). The Honmoku clay (w_L of 108.0% and w_P of 47.9%) was stabilized with a cement factor of 100 kg/m³ and cured for 14 days in a laboratory. The consistency of the stabilized soil measured at 14 days' curing is shown as black markers in the figure. The white markers in the figure show the measured values on the stabilized soil disturbed and compacted with additional cement mixing, which will be described later in Section 4. The figure shows that both the liquid limit, w_L, and the plastic limit, w_P, increase by the stabilization, which is different from the phenomenon shown in Figure 3.1. As the plastic limit increases more than the liquid limit, the plasticity index, I_p, is decreased by the stabilization.

3.2 Mechanical properties (strength characteristics)

3.2.1 Stress – strain curve

Figure 3.12 shows the stress–strain curves of soil field cement-stabilized soil by the pneumatic flow mixing method (Ministry of Transport, The Fifth District Port Construction Bureau, 1999). In the test, marine clay excavated at Nagoya Port

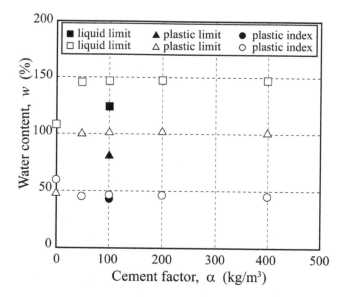

Figure 3.11 The change of consistency brought about by cement stabilization (Watabe & Tanaka, 2012).

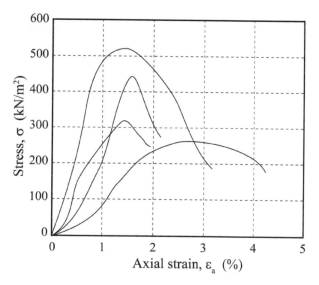

Figure 3.12 Stress – strain curves on cement-stabilized soils (Ministry of Transport, The Fifth District Port Construction Bureau, 1999).

(w_L of 74.4% and w_P of 33.0%) was stabilized with ordinary Portland cement and cured in the field. The stabilized soils were trimmed to 50 mm in diameter and 100 mm in height after core sampling and subjected to the unconfined compression test. The compressive stress, σ, increases rapidly with the axial strain to a peak strength which is

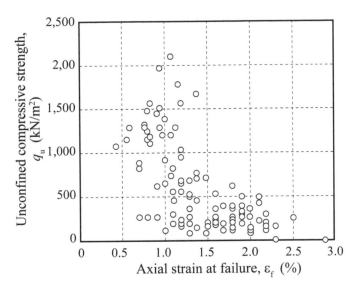

Figure 3.13 Strain at failure of cement-stabilized soils (Ministry of Transport, The Fifth District Port Construction Bureau, 1999).

followed by quick decrease in σ. The brittle behavior with a large peak strength, a small strain at failure and small residual strength becomes more dominant with the curing period.

3.2.2 Strain at failure

As shown in Figure 3.12, the axial strain at the failure of stabilized soil is quite small compared to that of the original soil. Figure 3.13 shows an example of the relationship between the axial strain at failure, ε_f, and the unconfined compressive strength, q_u, of field cement-stabilized soils (Ministry of Transport, The Fifth District Port Construction Bureau, 1999). In the tests, marine clay excavated at Nagoya Port (w_L of 74.4% and w_P of 33.0%) was stabilized with ordinary Portland cement and cured in the field. The stabilized soils were trimmed to 50 mm in diameter and 100 mm in height after core sampling and subjected to the unconfined compression test. The figure shows that the magnitude of axial strain at failure, ε_f, is of the order of 0.5 to 2.5% and considerably smaller than that of unstabilized clay. The ε_f decreases rapidly with the unconfined compressive strength, q_u. Similar test results can be seen for lime-stabilized soils as with cement-stabilized soils (e.g. Terashi et al., 1980).

3.2.3 Internal friction angle and undrained shear strength

Figure 3.14 shows the relationship between the consolidation pressure and the undrained shear strength, c_u, obtained in isotropically consolidated undrained shear (CIU) tests (Terashi et al., 1980). In Cases 1 to 3, the Kawasaki clay (w_L of 87.7% and w_P of 39.7%), having an initial water content of about 120%, was stabilized

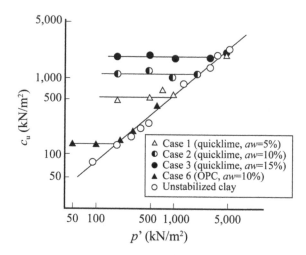

Figure 3.14 The relationship between the consolidation pressure and undrained shear strength (Terashi et al., 1980).

with quicklime with aw of 5, 10 and 15%, respectively. In Case 6, the Kawasaki clay, having an initial water content of about 200%, was stabilized with ordinary Portland cement with aw of 10%. In the figure, the test data on the unstabilized soil are also plotted by an open circle. The figure shows that the undrained shear strength, c_u, of the stabilized soil is larger than that of the unstabilized soil, and almost constant, as long as the consolidation pressure is low. But when the consolidation pressure exceeds the consolidation yield pressure (pseudo pre-consolidation pressure), p_y, the undrained shear strength increases with increasing consolidation pressure. This phenomenon can be seen irrespective of type and amount of binder. The increasing ratio in the c_u of the stabilized soil is almost same as that of the unstabilized soil. According to the figure, the internal friction angle, ϕ', of stabilized soil is almost zero as long as the consolidation pressure is lower than the consolidation yield pressure and the same as that of the unstabilized soil when the consolidation pressure is higher than the consolidation yield pressure.

Figure 3.15 shows the Mohr–Coulomb stress circles on the stabilized soil when tested for undrained shear strength (Ministry of Transport, The Fifth District Port Construction Bureau, 1999). The stress circle is almost same size as long as the confining pressure is lower than the cohesion of the stabilized soil.

Figure 3.16 shows the undrained shear strength of the cement-stabilized soil measured by various shear tests (Watabe et al., 2001). In the test, the Tachibana Bay clay, having an initial water content of $1.6 \times w_L$ was stabilized with ordinary Portland cement of $50\,\mathrm{kg/m^3}$. The stabilized soils were subjected to various shear tests: the unconfined compression test ($q_u/2$); the isotropically consolidated and undrained compression test (CIU); the anisotropically (K_0) consolidated and undrained compression test (CAUC); and the anisotropically (K_0) consolidated and undrained extension test (CAUE). In the case of the unconfined compression test, the undrained shear strength

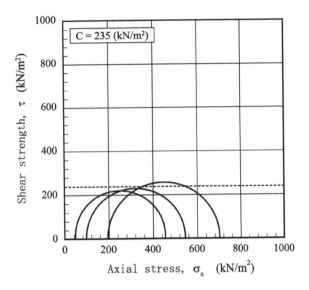

Figure 3.15 Mohr–Coulomb stress circles on cement-stabilized soil tested for undrained shear strength (Ministry of Transport, The Fifth District Port Construction Bureau, 1999).

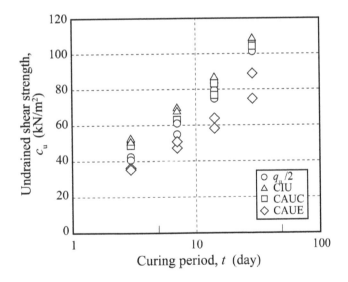

Figure 3.16 The undrained shear strength of cement-stabilized Tachibana clay measured by various testing methods (Watabe et al., 2001).

was assumed as $q_u/2$. The undrained shear strength is influenced by the testing method, where the largest value is obtained in the CIU test and the smallest is in the CAUE test. In general, the strength increases almost linearly with the logarithm of the curing period, irrespective of the testing method.

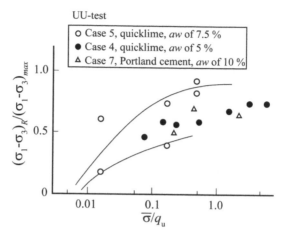

Figure 3.17 The relationship between residual strength and confining pressure (Terashi et al., 1980).

3.2.4 Residual strength

As already shown in Figures 3.4 and 3.12, the stress-strain curves and residual strength of stabilized soils are considerably influenced by the confining pressure, which show that the deviator stress, $\sigma_1 - \sigma_3$, sharply decreases after the peak in the case of unconfined compression, but the reduction in the deviator stress becomes smaller with the confining pressure, σ'_c. Figure 3.17 shows the relationship between the strength ratio of the residual strength against the peak strength, $(\sigma_1 - \sigma_3)_R/(\sigma_1 - \sigma_3)_{max}$ and the confining pressure ratio, $\overline{\sigma}/q_u$, which is obtained in unconsolidated undrained compression (UU) tests on the quicklime- and cement-stabilized clays having q_u value of 600 to 1,300 kN/m² (Terashi et al., 1980). The strength ratio increases with the confining pressure ratio, and is about 0.5 to 0.8 for the confining pressure ratio, $\overline{\sigma}/q_u$, exceeding about 0.1, irrespective of the type of binder (Terashi et al., 1980). A similar phenomenon was found where the residual strength of stabilized soil is about 0.8 of the unconfined compressive strength, q_u, even under a small confining pressure of the order of a couple of percentage of q_u (Tatsuoka & Kobayashi, 1983).

3.2.5 Modulus of elasticity (Young's modulus)

The modulus of elasticity of field cement-stabilized soils is plotted in Figure 3.18 against the unconfined compressive strength, q_u (Ministry of Transport, The Fifth District Port Construction Bureau, 1999). Marine clay excavated at Nagoya Port (w_L of 74.4% and w_P of 33.0%) was stabilized with ordinary Portland cement and cured in the field. The stabilized soils after core sampling were trimmed to 50 mm in diameter and 100 mm in height and subjected to the unconfined compression test. The modulus of elasticity, E_{50}, is defined by the secant modulus of elasticity in a stress–strain curve at a half of unconfined compressive strength, q_u. Although there is a lot of scatter in the measured data, the magnitude of E_{50} increases with the q_u and can be formulated as $E_{50} = 50$ to $300 \times q_u$.

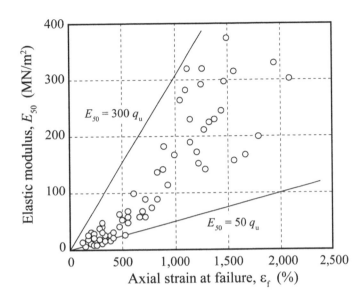

Figure 3.18 The modulus of elasticity, E_{50}, of laboratory cement-stabilized soils (Ministry of Transport, The Fifth District Port Construction Bureau, 1999).

Similar relationships were obtained on the laboratory lime-stabilized soils and laboratory cement-stabilized soils (e.g. Terashi et al., 1977, Niina et al., 1981). According to them, the E_{50} almost linearly increases with the q_u and the increasing ratio becomes larger with the q_u: $E_{50} = 50$ to $300 \times q_u$ for $q_u <$ about 2,000 kN/m², and $E_{50} = 350$ to $1,000 \times q_u$ for $q_u >$ about 2,000 kN/m².

3.2.6 Poisson's ratio

The Poisson's ratio, μ, of field cement-stabilized soils, is shown in Figure 3.19 against the unconfined compressive strength, q_u, in which the unconfined compression tests were carried out on small scale specimens of 50 mm in diameter (Niina et al., 1977). In the tests, the Shinagawa clay (w_L of 77.9% and w_P of 32.5%) was stabilized with quicklime, hydrated lime or cement by the deep mixing method (Kitazume & Terashi, 2013). The Poisson's ratio was obtained from the measured longitudinal and radial strains in the unconfined compression tests, and those for the shear stress lower than the 70% of q_u are plotted in the figure. Although there is a relatively large scatter in the test data, the Poisson's ratio is around 0.28 to 0.45 irrespective of the unconfined compressive strength, q_u. Accumulated researches show the Poisson's ratio is around about 0.2 to 0.45 irrespective of soil type, soil strength, size of soil specimen and laboratory and field stabilized soils (Hirade et al., 1995, The Building Center of Japan, 1997).

3.2.7 Dynamic property

Figure 3.20 shows the relationship between the initial shear modulus, G_0, at the shear strain of 10^{-6} and the unconfined compressive strength, q_u, of cement-stabilized

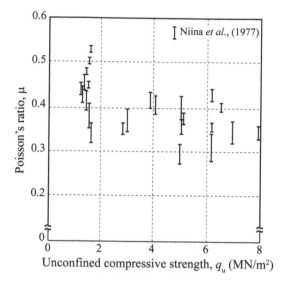

Figure 3.19 Poisson's ratio of *field* cement-stabilized soils (Niina et al., 1977).

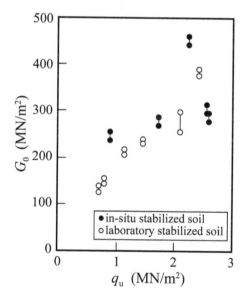

Figure 3.20 The relationship between the initial shear modulus and q_u of cement-stabilized soils (Tanaka & Terashi, 1986).

clays (Tanaka & Terashi, 1986). For the laboratory stabilized soil, clay excavated at Kawasaki Port (w_L of 88% and w_P of 44%), having an initial water content of 100 to 150%, was stabilized with ordinary Portland cement with aw of 10 to 25%, and the stabilized soils were subjected to the resonant column test. For the field stabilized soil, clay at Sakai Port (w_L of 93.3% and w_P of 27.3%) was stabilized *in-situ* by the deep mixing method (Kitazume & Terashi, 2013), where ordinary Portland cement of about 130 kg/m^3 was mixed and the mixture was cured *in-situ*. At 140 days after the execution, the stabilized soil was core sampled and trimmed for the test. The initial shear modulus, G_0, almost linearly increases with the q_u and can be formulated as $G_0 = 80$ to $200 \times q_u$, irrespective of the laboratory and field stabilized soils.

Figure 3.21 shows the shear modulus ratio, G/G_0, and the damping ratio, h_{eq}, against the shear strain, γ, of the field cement-stabilized soil by the pneumatic flow mixing method (Ministry of Transport, The Fifth District Port Construction Bureau, 1999). Marine clay excavated at Nagoya Port (w_L of 74.% and w_P of 31.6%) was prepared to the initial water content of 107% and stabilized with blast furnace slag cement type B with a cement factor, α, of 60 kg/m^3 and placed on land. The stabilized soil was core sampled at 28 days' curing. In the case of the confining pressure of 50 kN/m^2 (Figure 3.21(a)), the shear modulus ratio, G/G_0, is almost constant irrespective of the shear strain, while the equivalent dumping ratio, h_{eq}, increases gradually. The shear modulus ratio decreases with the shear strain in the cases of the confining pressure of 100 and 200 kN/m^2 (Figures 3.21(b) and 3.21(c)), while the dumping ratio increases to about 10% with the shear strain. In the figure, an example of test data on an ordinary clay having the plasticity index, I_p, exceeding 30 is plotted by broken lines. The shear modulus ratio and dumping ratio of the stabilized soil are similar to those of the ordinary clay.

The dynamic property of laboratory stabilized sand was also reported by Shibuya et al. (1992) and Enami et al. (1993).

3.2.8 Creep strength

Figure 3.22 shows the relationship between the strain rate, $\dot{\varepsilon}$, and the loading period of cement-stabilized clay (Terashi et al., 1983). The Kawasaki clay (w_L of 88% and w_P of 40%), having an initial water content of 150 or 200%, was stabilized with ordinary Portland cement with a cement content, aw, of 15 or 20%. The specimen of 50 mm in diameter and 100 mm in height was subjected to the constant compressive pressure, q_{cr}, whose magnitude was changed from 0.52 to $0.91 \times q_u$. The strain rate, $\dot{\varepsilon}$, decreases almost linearly with the time duration on the double-logarithmic graph. The decreasing phenomenon in $\dot{\varepsilon}$ is almost constant irrespective of the load intensity, q_{cr}/q_u. The figure shows that the stabilized soil subjected to the vertical load with q_{cr}/q_u of 0.91, exhibits creep failure at around 1 minute after loading, but the stabilized soil does not fail as long as the load intensity is lower than about $0.8 \times q_u$.

3.2.9 Cyclic strength

Figure 3.23 shows the relationship between the axial compressive strain, ε, and the number of loading cycles, N, on the cement-stabilized soils (Ministry of Transport, The Fifth District Port Construction Bureau, 1999). In the tests, marine clay excavated at Nagoya Port (w_L of 74.% and w_P of 31.6%) was prepared to an initial water content of 107% and stabilized in the field with blast furnace slag cement type B with cement factor, α, of 60 kg/m^3 by the pneumatic flow mixing method. The stabilized soil cured in the field was core sampled at 28 days. The samples were subjected to

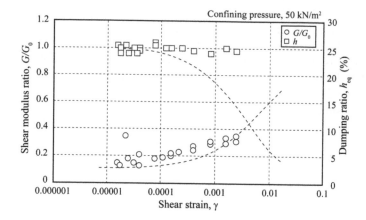

(a) Confining pressure, 50 kN/m^2.

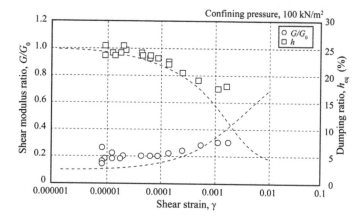

(b) Confining pressure, 100 kN/m^2.

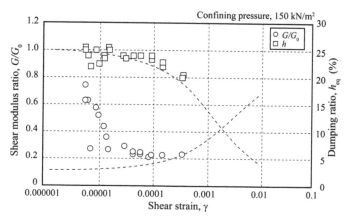

(c) Confining pressure, 150 kN/m^2.

Figure 3.21 The shear modulus ratio and damping ratio against the shear strain of field cement-stabilized soil (Ministry of Transport, The Fifth District Port Construction Bureau, 1999).

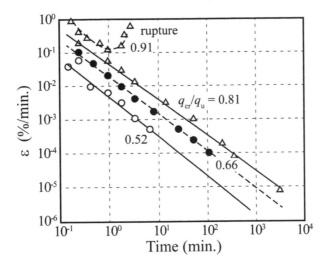

Figure 3.22 The relationship between strain rate and the loading period in the creep test (Terashi et al., 1983).

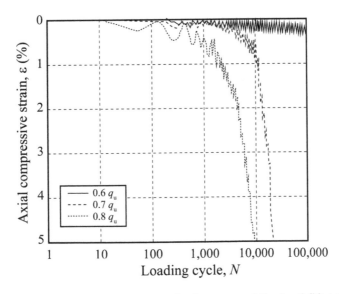

Figure 3.23 The effect of cyclic loading on strength of cement-stabilized soil (Ministry of Transport, The Fifth District Port Construction Bureau, 1999).

the cyclic loading of 1 Hz with the maximum deviator stress of $0.6 \times q_u$, $0.7 \times q_u$ and $0.8 \times q_u$ (q_u is $307 \, kN/m^2$) while the confining pressure was kept constant at $50 \, kN/m^2$. The figure shows that the axial compressive strain remains lower than about 0.4% in the case of $0.6 \times q_u$ even with the number of cyclic loading being 100,000. In the cases of the compressive stress of $0.8 \times q_u$ and $0.7 \times q_u$, on the other hand,

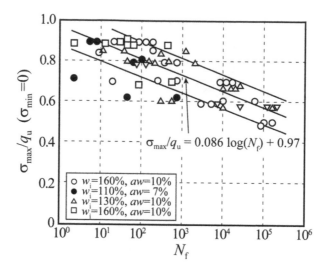

Figure 3.24 The relationship between number of cyclic loading until failure and magnitude of loading (Terashi et al., 1983).

the axial compressive strain rapidly increases to failure when the number of cyclic loading exceeds about 3,000 and 10,000, respectively.

Similar tests were carried out by Terashi et al. (1983) and Kudo et al. (1996). Figure 3.24 shows one of the relationships between the normalized cyclic loading stress, σ_{max}/q_u, and the number of cyclic loading at failure, N_f, on cement-stabilized soils (Terashi et al., 1983). In the tests, the Kawasaki clay, having an initial water content of 200%, was stabilized with ordinary Portland cement with α of 15%, whose unconfined compressive strength was measured to be about 470 kN/m^2. The stabilized soil was subjected to the cyclic loading with various compressive stresses, σ_{max}, while the minimum pressure, σ_{min}, was kept constant at 0 kN/m^2. The number of cyclic loading at failure, N_f, in the logarithmic scale decreases almost linearly with the σ_{max}/q_u. The number of cyclic loading at failure, N_f, for the $\sigma_{max}/q_u = 0.6$, 0.7 and 0.8 are around 2,000 to 400,000, 150 to 30,000, and 100 to 2,000, respectively. These are similar to, but smaller than, those phenomena in Figure 3.23.

3.2.10 Tensile and bending strengths

The tensile and bending strengths of stabilized soil have been evaluated by various tests: the split tension test (Brazilian tension test, indirect tension test), the simple tension test, and the bending test. The influence of the testing method on the tensile and bending strengths were discussed in detail by Saitoh et al. (1996) and Namikawa & Koseki (2007).

The tensile and bending strengths of stabilized soil were evaluated by the split tension tests and bending tests (Terashi et al., 1980). In the tests, the Kawasaki clay (w_L of 87.8% and w_P of 39.7%), having various initial water contents, w_i, was stabilized with either quicklime or ordinary Portland cement to form a disc-shaped specimen

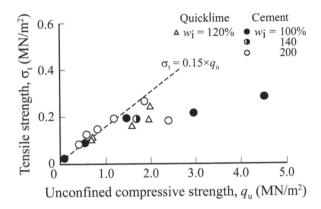

(a) Measured by split tension tests.

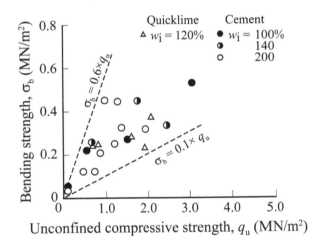

(b) Measured by bending tests.

Figure 3.25 Tensile strength of laboratory stabilized soils (Terashi et al., 1980).

of 100 mm in diameter and 50 mm in height for the split tension test and a beam with rectangular cross section of 50 mm in width, 50 mm in height and 250 mm in length for the bending test. Figure 3.25(a) shows the relationship between the tensile strength, σ_t, and the unconfined compressive strength, q_u. The figure shows the tensile strength increases almost linearly with the unconfined compressive strength, q_u, irrespective of the type, amount of binder and the initial water content of original soil, but its increment ratio becomes smaller when the q_u exceeds about 2 MN/m². The tensile strength is about 0.15 of the unconfined compressive strength, q_u. Figure 3.25(b) shows the relationship between the tensile strength evaluated by the bending test, σ_b, and the unconfined compressive strength, q_u. The figure shows the bending strength is around 0.1 to 0.6 of q_u irrespective of the type of binder and the initial water content of original soil.

3.2.11   *Long-term strength*

The accumulated data on the long-term strength in years or decades has revealed that the long-term strength is influenced by the exposed condition. Figure 3.26(a) shows an example of the long-term strength of stabilized soil under various curing conditions (Terashi et al., 1983). The Kawasaki clay (w_L of 88% and w_P of 40%), having an initial water content of 110%, was stabilized with ordinary Portland cement with a cement content, aw, of 10%. The stabilized soil was placed in a plastic mold of

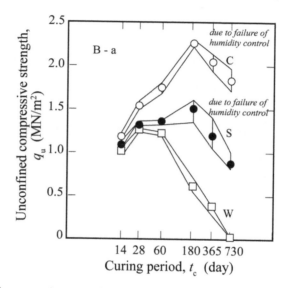

(a) The unconfined compressive strength for different curing conditions (remolded at 24 hours after mixing).

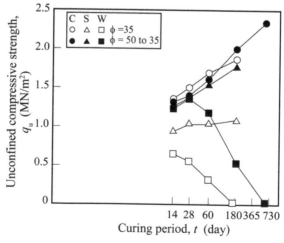

(b) Unconfined compressive strength of the centre portion of the soil sample.

Figure 3.26 The unconfined compressive strength for different curing conditions (Terashi et al., 1983).

50 mm in diameter and 100 mm in height, and retrieved from the mold at 2 hours and 24 hours after mixing, when its unconfined compressive strength was about 60 and 640 kN/m², respectively. The stabilized soils were cured either in a chamber with a constant temperature of 20 °C and 100% relative humidity, embedded in the Kawasaki clay with a 120% water content, or submerged in seawater. Figure 3.26(a) shows one of the test results, where the letters, C, S and W correspond to be cured in the chamber, embedded in the soil and submerged in seawater respectively. In the case of the samples cured in the chamber, C, the strengths increase almost linearly with the curing period in a logarithmic scale (the strength decrease after 180 days was due to the failure of humidity control). In the cases of cured in the soil, S, the strengths increase at the beginning to peak strengths at 180 days (the strength decrease after 180 days was due to the failure of humidity control). In the cases of cured in the seawater, W, the strengths increase at beginning to peak strengths at 28 days and then they decrease rapidly with the curing period.

Figure 3.26(b) shows the other test result on the long-term strength under various curing conditions in order to show the influence of deterioration on the periphery of a soil sample (Terashi et al., 1983). The term, $\phi = 35$, in the graph legend indicates the stabilized soil sample of 35 mm in diameter was cured and subjected to an unconfined compression test. The term, $\phi = 50$ to 35, indicates that the soil sample of 50 mm in diameter was cured and its periphery was trimmed out to make a sample of 35 mm in diameter for the unconfined compression test. The latter case evaluates the strength of the core portion of stabilized soil. The figure shows the unconfined compressive strength of both the core and peripheral portions increase in the case of the sample cured in the chamber, C. In the case of the sample cured in the soil, S, the strength of the core portion increases at almost the same ratio as the C sample, while the strength of $S - \phi = 35$ sample increases slowly with the curing period. This means the peripheral portion of stabilized soil was deteriorated and lost the strength. In the case of the sample cured in seawater, W, the strength decreases rapidly in the both samples, $\phi = 35$ and $\phi = 50$ to 35, which means the soil sample deteriorated very severely and rapidly not only in the peripheral portion, but also the core portion of the soil.

The test data presented above have revealed that two aspects proceed at the core portion and periphery portion simultaneously as shown in Figure 3.27. One is the strength increase with time at the core portion of stabilized soil which is negligibly influenced by the surrounding conditions and the other is the possible strength decrease

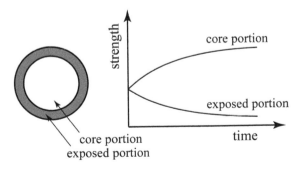

Figure 3.27 Image of long-term strength of stabilized soil.

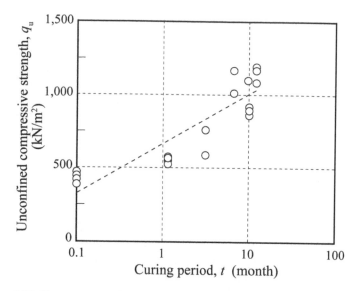

Figure 3.28 The relationship between q_u and elapsed time (Kitazume et al., 2003).

with time at the peripheral (exposed) portion of the stabilized soil column due to the deterioration that is influenced by the surrounding exposure conditions.

3.2.11.1   Long-term strength of core portion

Figure 3.28 shows an example of the long-term strength of stabilized soil with the curing period (Kitazume et al., 2003). The figure shows the relationship between the unconfined compressive strength, q_u, and the curing period on laboratory stabilized soil. In the tests, the Kawasaki marine clay (w_L of 83.4% and w_P of 38.6%) was thoroughly mixed with an initial water content of 160%, and was mixed with ordinary Portland cement with a cement content, aw, of 30% to make soil samples of 50 mm in diameter and 100 mm in height. The specimen was wrapped in a high polymer film and stored in the humid chamber at a temperature of 20°C and 95% relative humidity, which corresponds to the core portion as already shown in Figure 3.28. Figure 3.29 clearly shows the unconfined compressive strength increases almost linearly with the logarithm of the curing period.

The long-term strength increase has also been studied on the *in-situ* stabilized soils by the deep mixing method and shallow mixing method (Niina et al., 1981; Terashi & Kitazume, 1992; Niigaki et al., 2000; Hayashi et al., 2003, 2004; Ikegami et al., 2002a, 2002b, 2003, 2005; Kitazume & Takahashi, 2009). In these studies the soil specimens were core sampled from the *in-situ* stabilized soil column and trimmed for an unconfined compression test, which can be approximated to a sort of core portion. Figure 3.29 shows the relationships between the unconfined compressive strength, q_u, and the elapsed time of more than 20 years, in which several types of soil were stabilized with various types and amounts of binder. The strength of stabilized soil is highly dependent upon the type of soil, and the type and amount of binder. However,

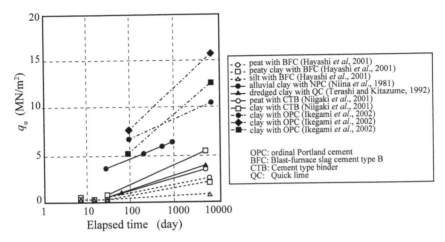

Figure 3.29 The relationship between q_u and the curing period of field situstabilized soils.

the strength of stabilized soil increases almost linearly with the logarithm of the curing period, irrespective of the type of soil, and the type and amount of binder.

The long-term strength of the stabilized soil produced by the pneumatic flow mixing method was also investigated in the Central Japan International Airport construction project (Chapter 4). As shown later in Figure 4.11, no clear change was found in the unconfined compressive strength and the water content of the stabilized soil up to 10 years' curing (Kitazume et al., 2006, Morikawa et al., 2012).

3.2.11.2 Long-term strength of the peripheral portion

(a) Strength profile within stabilized soil

Figure 3.30 shows the strength profile of laboratory cement-stabilized soil along the distance from the exposed surface (Hara et al., 2013). In the test, the Ariake clay (w_L of 158.1% and w_P of 51.4%), having an initial water content of $1.5 \times w_L$, was stabilized with ordinary Portland cement with a cement factor, α, of 50, 70, and 100 kg/m^3. After four weeks' curing in a controlled-environment chamber, one surface of the cylindrical-shaped specimen was exposed to artificial seawater with a concentration degree of 20 g/L of sodium chloride (NaCl). The seawater was either not exchanged, or exchanged every week, during the test period. The unconfined compressive strengths at 28 days' curing, q_{u28}, for α of 50, 70 and 100 kg/m^3 were 239, 657 and 1,221 kN/m^2, respectively. At the prescribed curing period, the strength profile of the stabilized soil was measured with the small size cone penetration test.

Figure 3.30 shows that the penetration resistance at 0 day (initial) shows the rapid increase with the penetration depth to a constant value at the depth of about 10 mm irrespective of the cement factor. The resistance at the deep portion increases with the curing period, which indicates the increase in the soil strength at the core portion. At the shallow depth portion, on the other hand, the penetration resistance is quite small irrespective of the cement factor and the test condition. The deterioration depth, which is defined as the depth where the penetration resistance increases quickly,

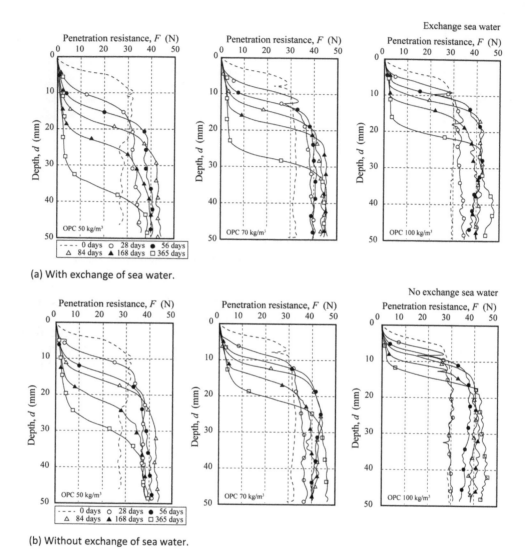

(a) With exchange of sea water.

(b) Without exchange of sea water.

Figure 3.30 Spatial strength distribution in samples of laboratory-produced stabilized soils (Hara et al., 2013).

becomes larger with the curing period. The deterioration depth becomes large with decreasing cement factor.

The effect of exposed conditions on the long-term strength was discussed in laboratory tests by Kitazume et al. (2003), where the Kawasaki clay (w_L of 83.4% and w_P of 38.6%) stabilized with ordinary Portland cement, was exposed to either tap water, seawater or clay. A specimen wrapped in sealant was also prepared and cured for reference. At the prescribed curing period, the strength profile of stabilized soil was measured by the needle penetration test. In the cases of exposure to tap water

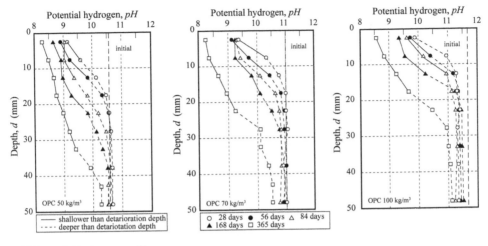

(a) Potential hydrogen profile.

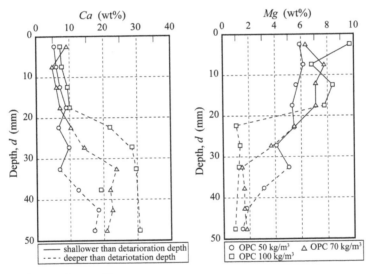

(b) Calcium and magnesium profiles.

Figure 3.31 Potential hydrogen, calcium and magnesium distributions in stabilized soil (Hara et al., 2013).

and seawater, the strength of stabilized soil close to the exposed surface decreased very rapidly and the deterioration progressed gradually inward with increasing curing period. However, in the case of exposure to clay, negligible strength decrease in the specimen was found, even after twelve months' exposure.

(b) Potential hydrogen, pH, calcium and magnesium profiles within stabilized soil
Figure 3.31(a) shows the potential hydrogen, pH, profile within the stabilized soil (Hara et al., 2013). The pH value is about 10.5 to 11.5 at the core portion, and increases

with the cement factor. The pH value at the periphery (deterioration portion) is smaller than that at the core portion, and is about 8 to 10 at the exposed surface. The extent of the low pH portion is similar to the penetration resistance as shown in Figure 3.31(a).

Figure 3.31(b) shows the calcium and magnesium distributions in the stabilized soil at 365 days' curing. The Ca^{2+} value at the deep portion shows a large value and is dependent on the cement factor, but the value at the shallow portion is small irrespective the cement factor, which indicates that the Ca^{2+} in the stabilized soil elutes to 5 to 10% with the curing period. The Mg^{2+} distribution shows the opposite phenomenon, where the Mg^{2+} value at the deep portion is small irrespective of the cement factor, but the value at the shallow portion is large and its magnitude is dependent on the cement factor. These phenomena are consistent with the penetration resistance profile as shown in Figure 3.31.

The long-term strength of *in-situ* stabilized soil was investigated in detail at Yokohama Port (Ikegami et al., 2002a, 2002b, 2003, 2005). The thick alluvial clay ground was improved by the wet method of deep mixing method (Kitazume & Terashi, 2013) with ordinary Portland cement of 180 kg/m³. Figure 3.32(a) shows the strength and calcium content distributions in the cement-stabilized soil after 20 years' curing in the ground (Ikegami et al., 2002a, 2002b). The horizontal axis of the figure is the horizontal distance from the exposed surface in logarithmic scale. The strength in terms of unconfined compressive strength shown in the upper half of the figure was estimated, based on the needle penetration test. The calcium content shown in the lower half of the figure was measured in sliced core samples by means of atomic adsorption spectrometry. The overall pattern of strength and calcium content distributions are in good agreement each other except for the large calcium content found at 5 to 10 mm from the exposed surface. Figure 3.32(b) shows the calcium content distribution across the exposed surface between stabilized soil and original soil (Ikegami et al, 2005). The calcium content in the stabilized soil decreases toward the exposed surface, and that in the original soil increases toward the exposed surface. The overall pattern of calcium content suggests that the calcium leaching from the stabilized soil to the original soil is the dominating phenomenon which caused the deterioration at the periphery.

(c) Deterioration depth

Figure 3.33 shows the relationship between the deterioration depth and curing period (Hara et al., 2013). The soil and mixing condition are the same as in Figure 3.30. The deterioration depth increases with the curing period, irrespective of the cement factor and test conditions. The deterioration speed is larger in the low cement factor soil than in the high cement factor soil.

Figure 3.34 compares the depth of deterioration with the curing period (Ikegami et al., 2002a, 2002b). In the figure, the *in-situ* stabilized soil by the deep mixing method at Daikoku Pier and the laboratory exposure test results by Terashi et al. (1983), Saitoh (1988), Kitazume et al. (2003) and Hayashi et al. (2004) are plotted together. The strength, q_{u28}, in the legend of the figure is the unconfined compressive strength of stabilized soil after 28 days' curing under the sealed condition. The progress of deterioration depth in logarithmic scale is almost linear to the curing period in logarithmic scale, and the gradients of all the test cases are about 1/2 irrespective

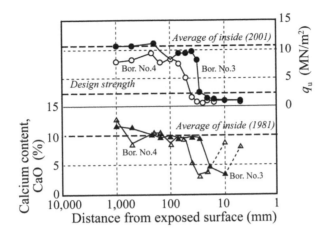

(a) Strength and calcium content distribution in stabilized soil.

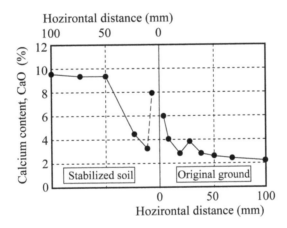

(b) Calcium content distribution in both stabilized soil and original soil.

Figure 3.32 The long-term strength and calcium content in *in-situ* stabilized soil (Ikegami et al., 2002a, 2002b, 2005).

of the strength of soil and exposure condition, which means the rate of deterioration is proportional to the square root of time. A similar relationship between the depth of deterioration and time was also obtained by a numerical simulation proposed by Nishida et al. (2003) that assumed ions' migration primarily based on diffusion by the Ca concentration gradient. Judging from the results of laboratory tests and the numerical analysis, it may be possible to predict long-term deterioration by extrapolation of the short-term result of the exposure test assuming the deterioration progress is in proportion to a square root of time. General tendency found in Figure 3.34 is that the larger the strength the smaller the depth of deterioration and seawater exposure gives rise to larger depth of deterioration compared to tap water exposure or contact to original soil.

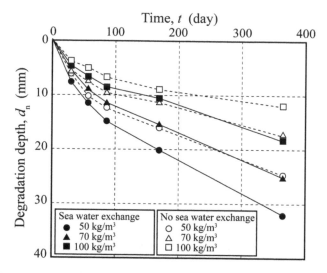

Figure 3.33 The relationship between deterioration depth and curing period (Hara et al., 2013).

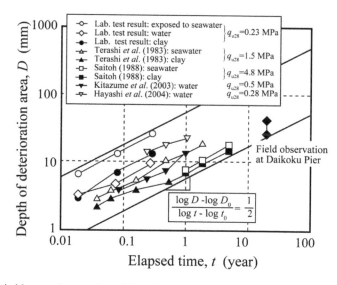

Figure 3.34 Measured extension of deterioration with elapsed time (Ikegami et al., 2005).

3.2.12 Coefficient of horizontal stress at rest

Figure 3.35 shows the coefficient of horizontal earth pressure and axial strain in an anisotropically (K_0) consolidation test (Coastal Development Institute of Technology, 2008). Marine clay at Nagoya Port (w_L of 74.9% and w_P of 31.6%) was prepared to have an initial water content of 107% and stabilized with blast furnace slag cement

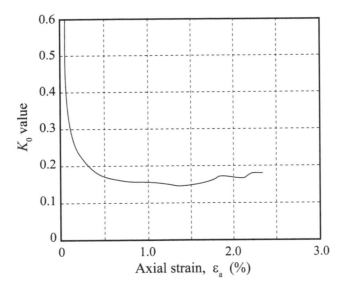

Figure 3.35 The coefficient of horizontal stress at rest (Coastal Development Institute of Technology, 2008).

with α of 60 kg/m³ by the pneumatic flow mixing method in the field. The stabilized soil was core sampled at 28 days' curing. In the laboratory test, the soil sample was consolidated isotropically with a consolidation pressure of 10 kN/m² at first, and then was consolidated anisotropically (K_0 consolidation) with an axial consolidation pressure of 200 kN/m². The coefficient of horizontal stress at rest, K_0 value obtained in the test is about 0.15.

3.3   Mechanical properties (consolidation characteristics)

3.3.1   Void ratio – consolidation pressure curve and consolidation yield pressure

Figure 3.36(a) shows the void ratio, e, and the consolidation pressure, p, of field cement-stabilized soils (Ministry of Transport, The Fifth District Port Construction Bureau, 1999). Marine clay excavated at Nagoya Port (w_L of 74.4% and w_P of 33.0%) was stabilized with ordinary Portland cement of various cement factors by the pneumatic flow mixing method and cured in the field. The stabilized soil, after core sampling, was trimmed to 60 mm in diameter and 20 mm in height and subjected to the oedometer test. For the large scale sample, the fresh stabilized soil was sampled at the outlet of the placement machine and molded to a consolidation cell of 300 mm in diameter and 100 mm in height and cured in a laboratory. The e-log p curves of the oedometer test on the stabilized soils with $C = 0$ kg/m³ (original soil) and $C = 38$ kg/m³ show an almost linear relationship without clear bending points, consolidation yield pressure. In the cases of the oedometer test with $C = 57$ and 68 kg/m³, a clear consolidation yield pressure can be seen and its magnitude is dependent on the cement factor.

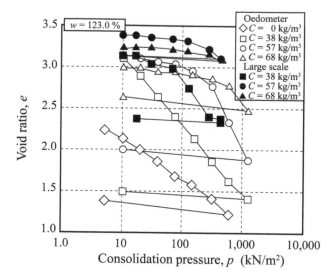

(a) *E*-log *p* curve.

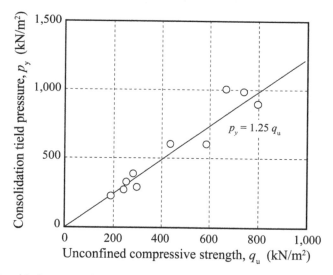

(b) Relationship between the consolidation yield pressure, p_y, and unconfined compressive strength, q_u.

Figure 3.36 The e-log *p* curve, and consolidation yield pressure (Ministry of Transport, The Fifth District Port Construction Bureau, 1999).

In the cases of the large scale samples, a clear consolidation yield pressure can be seen even in the stabilized soil with $C = 38 \, kg/m^3$. A similar phenomenon was obtained with several types of soil and binder (Terashi et al., 1980, Kawasaki et al., 1978, Takahashi & Kitazume, 2004).

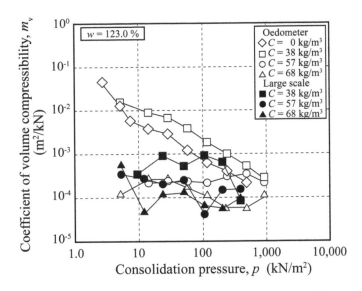

Figure 3.37 The relationship between the coefficient of volume compressibility and consolidation pressure (Ministry of Transport, The Fifth District Port Construction Bureau, 1999).

Figure 3.36(b) shows the relationship between the consolidation yield pressure, p_y, and the unconfined compressive strength, q_u, of the Nagoya Port clay (w_L of 74.4% and w_P of 33.0%) stabilized with ordinary Portland cement (Ministry of Transport, The Fifth District Port Construction Bureau, 1999). The figure shows that the consolidation yield pressure, p_y, has a linear relationship with the unconfined compressive strength, q_u. The ratio of p_y/q_u of the stabilized soils is 1.25. A similar phenomenon was obtained on several types of soil and binder, where the ratio of p_y/q_u is about 1.25 to 1.55 irrespective of the types of original soil and binder (Terashi et al., 1980, Kawasaki et al., 1978, Takahashi & Kitazume, 2004).

3.3.2   Coefficient of volume compressibility

Figure 3.37 shows the coefficient of volume compressibility of the stabilized soil, m_v (Ministry of Transport, The Fifth District Port Construction Bureau, 1999). Marine clay excavated at Nagoya Port (w_L of 74.4% and w_P of 33.0%) was stabilized with ordinary Portland cement of various cement factors and cured in the field. The testing procedure and conditions are the same as those in Figure 3.37. The figure indicates that the coefficient of volume compressibility of the stabilized soils, m_{vs}, slightly decreases with the consolidation pressure, p, except in the case of the oedometer test on the sample of stabilized soil with $C = 38$ kg/m³. In the case of the oedometer test of $C = 38$ kg/m³, the m_v decreases rapidly with the consolidation pressure and its decrease is very similar to that of the original soil.

Figure 3.38 shows the other relationship between the coefficient of volume compressibility of stabilized soils, m_{vs}, and the consolidation pressure, p (Terashi

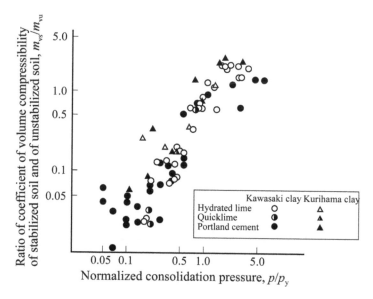

Figure 3.38 The relationship between the coefficient of volume compressibility and consolidation pressure on laboratory stabilized soils (Terashi et al., 1980).

et al., 1980). In the tests, a series of one-dimensional consolidation tests on the stabilized and unstabilized soils were carried out under a wide range of consolidation pressures. Two marine clays, the Kawasaki clay (w_L of 87.7% and w_P of 39.7%) and the Kurihama clay (w_L of 70.9% and w_P of 30.8%), were stabilized either with hydrated lime, quicklime or ordinary Portland cement. The size of soil samples was 20 mm in thickness and 60 mm in diameter. The coefficient of volume compressibility, m_{vs}, was normalized by that of the original soil, m_{vu}, at the same consolidation pressure and is shown on the vertical axis. The consolidation pressure, p, is also normalized by the consolidation yield pressure of the stabilized soil, p_y, and shown on the horizontal axis. The figure shows the ratio of m_{vs}/m_{vu} is 0.01 to 0.1 as long as the normalized consolidation pressure, p/p_y is around 0.1, but the m_{vs}/m_{vu} increases to unity at the p/p_y of 1.

3.3.3 Coefficient of consolidation

Figure 3.39 shows the relationship between the coefficient of consolidation of stabilized soil, c_{vs}, and the consolidation pressure, p (Ministry of Transport, The Fifth District Port Construction Bureau, 1999). The test conditions are the same as those in Figure 3.37. The c_{vs} of the stabilized soil decreases gradually with the consolidation pressure, while the c_{vu} of the unstabilized soil increases gradually. The figure shows the c_{vs} value is about 10 to 100 time larger than that of unstabilized soil as long as the consolidation pressure is small, but the c_v becomes almost the same as that of the stabilized soil when the consolidation pressure becomes large, of the order of 1,000 kN/m².

Figure 3.40 shows the relationship between the coefficient of consolidation of the stabilized clays, c_{vs}, and the consolidation pressure, p (Terashi et al., 1980). The test

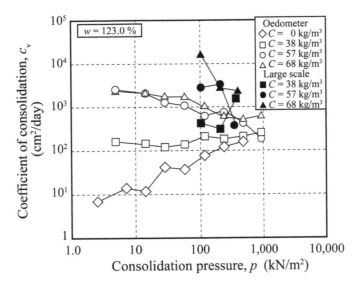

Figure 3.39 The relationship between the coefficient of consolidation and consolidation pressure on laboratory stabilized soils (Ministry of Transport, The Fifth District Port Construction Bureau, 1999).

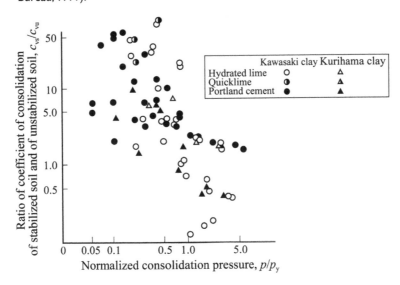

Figure 3.40 The relationship between the coefficient of consolidation and consolidation pressure on laboratory stabilized soils (Terashi et al., 1980).

conditions are the same as those in Figure 3.36. Similar to Figure 3.38, the coefficient of consolidation of the stabilized soil, c_{vs}, is normalized by that of the unstabilized soil, c_{vu}, at the same consolidation pressure, and the consolidation pressure, p, is also normalised by the consolidation yield pressure of the stabilized soil, p_y. The figure shows the ratio of c_{vs}/c_{vu} is 10 to 100 as long as the normalised consolidation pressure,

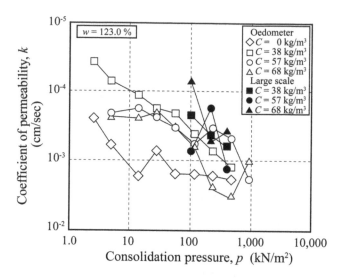

Figure 3.41 The relationship between permeability and consolidation pressure of laboratory stabilized soils (Ministry of Transport, The Fifth District Port Construction Bureau, 1999).

p/p_y is around 0.1, in a sort of over-consolidated condition, but the c_{vs}/c_{vu} approaches to unity at the p/p_y of 1, c_{vs}/c_{vu}, 1 when p/p_y exceeds 1.

3.3.4 Coefficient of permeability

Figure 3.41 shows the relationship between the coefficient of permeability of stabilized soil, k, and the consolidation pressure, p, which was measured in the oedometer test (Ministry of Transport, The Fifth District Port Construction Bureau, 1999). The soil preparation is the same as that in Figure 3.36. The k value of the unstabilized soil is the order of around 10^{-8} to 10^{-7} cm/s and slightly decreases with increasing consolidation pressure. In the cases of the stabilized soils, the k value is larger than that of unstabilized soil when the consolidation pressure is small, but the k decreases with the consolidation pressure and becomes almost same as the unstabilized soil when the consolidation pressure becomes large, of the order of 1,000 kN/m².

Figure 3.42 shows the coefficient of permeability of the stabilized Kawasaki clay (w_L of 87.7% and w_P of 39.7%) with ordinary Portland cement with the cement content, aw of 5, 10 and 15%, in which the coefficient of permeability is plotted against the water content of the stabilized soils (Terashi et al., 1983). In the tests, the stabilized soil specimens, 20 mm in height and 50 mm in thickness, were subjected to the constant head permeability tests. The figure shows that the coefficient of permeability is dependent upon the water content of stabilized soil and the amount of cement. The coefficient of permeability of the stabilized soil decreases with decreasing water content and with increasing amount of cement.

Figure 3.43 shows the relationship between the coefficient of permeability and the strength of laboratory stabilized soil (Terashi et al., 1983). The coefficient of permeability of the stabilized soil decreases exponentially with increasing strength, q_u.

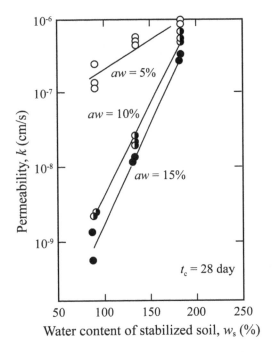

Figure 3.42 The relationship between the coefficient of permeability and water content of cement-stabilized soils (Terashi et al., 1983).

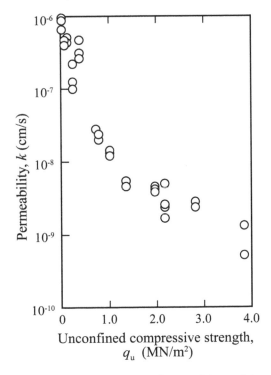

Figure 3.43 The relationship between the coefficient of permeability and the unconfined compressive strength of stabilized soil (Terashi et al., 1983).

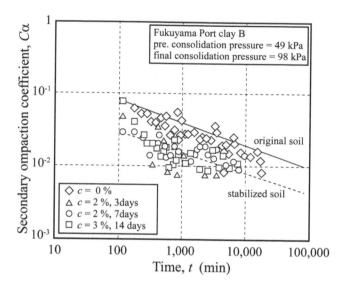

Figure 3.44 Secondary compression coefficient of stabilized soil (afterUdaka et al., 2013).

The relationship between the coefficient of permeability and the void ratio was discussed by Takahahi & Kitazume (2004) and Onitsuka et al. (2003). The influence of grain size distribution on the coefficient of permeability was discussed by Miura et al. (2004).

3.3.5 Secondary compression

Figure 3.44 shows the secondary compression coefficient, C_α, with the consolidation time (Udaka et al., 2013). In the test, the Fukuyama Port clay was mixed throughout to an initial water content of $1.5 \times w_L$ and mixed with ordinary Portland cement with a cement content, aw, of 7.0 or 8.0%, while the W/C ratio of cement slurry was 100%. The stabilized soil mixture was poured in the consolidation mold, 120 mm in diameter and 250 mm in height, and subjected to the pre-consolidation one dimensionally under the consolidation pressure of 49 kN/m² for 3, 7 or 14 days. The soil sample was finally subjected to the consolidation pressure of 98 kN/m² for the long-term consolidation test. The secondary compression coefficients, C_α, of the stabilized soil and original soil decrease almost linearly with the logarithm of consolidation time, and the coefficient of the stabilized soil is about 1/3 to 1/2 of the original soil. The figure also shows the stabilized soil with aw of 7% and subjected to 3 days' pre-consolidation shows the lowest C_α value, which may be due to the hydration effect during the long-term consolidation.

3.4 Environmental properties

3.4.1 Elution of contaminant

The Soil Contamination Countermeasures Act (Environment Agency, 1975, 2005) was enforced by the Ministry of Environment of the Japanese government in 2005, in order

Table 3.1 Soil elution criterion and second elution criterion designated by the Soil Contamination Countermeasures Act, 1975, 2005.

Hazardous substance	Soil elution criterion (mg/l)	Second elution criterion (mg/l)
Cadmium	0.01	0.3
Lead	0.01	0.3
Hexavalent chromium	0.05	1.5
Arsenic	0.01	0.3
Mercury	0.0005	0.005
Selenium	0.01	0.3
Fluorine	0.8	24
Boron	1	30
Trichloroethylene	0.01	0.3

to facilitate the implementation of countermeasures against soil contamination and measures for the prevention of harmful effects on human health, and thereby to protect the health of its citizens. In the Act, 26 chemical substances, including lead, arsenic, and trichloroethylene are a 'Designated Hazardous Substance' which can have harmful effects on human health (Table 3.1). The Act designates that not only natural soils but also stabilized soils shall be subjected to soil contamination investigation to measure the content and elution amount of the substances and report them to the governor. Four regulated values are designated in the Act, of which 'soil elution criterion' and 'second elution criterion' are the critical concerns for excavation and filling soils. The former is designated by the Minister of the Environment for the 'Designated Areas'. When the situation of contamination of the soil at a site by a Designated Hazardous Substance does not conform to the criteria, the prefectural governor shall designate an area covering such a site as an area contaminated by the Designated Hazardous Substance. The soils in the 'Designated Areas' should be stabilized by *in-situ* insolubility, *in-situ* confinement or confinement by an impermeable wall.

The effects of stabilization on the leaching of hazardous substances were investigated by a series of laboratory leaching tests, where five soils artificially contaminated by the eight chemical reagents designated as 'Designated Hazardous Substances' were prepared and stabilized with a cement-based special binder (Kaneshiro et al., 2006). The properties of the soils and the chemical reagents are summarised in Tables 3.2 and 3.3, respectively. The stabilized soils were prepared by the procedure specified by the Japan Cement Association (JCAS L-1: 2006), which is almost the same as the Japanese Geotechnical Society standard (2009). After 7 days' curing, the leaching tests were carried out on the specimens according to the testing procedure specified by the environmental Quality Standards (Environment Agency, 1975), where the stabilized soil was crushed into pieces, sieved through a 2 mm sieve and dried naturally in advance of the tests.

Figure 3.45 shows the leaching test results on the eight hazardous substances shown in Table 3.3. For cadmium leaching from the stabilized soil (Figure 3.45(a)), the amount of cadmium leaching quickly decreases with an increasing cement factor and becomes lower than the detection limit for all the stabilized soils. For lead leaching

Table 3.2 Physical properties of the stabilized soils tested for leaching of hazardous substances (Kaneshiro et al., 2006).

	Water content (%)	Density (g/cm^3)	pH	Particle size distribution (%)			Classification
				Gravel	Sand	Fine	
Sand (1)	34	1.849	3.88	0.0	53.4	46.6	SF
Sand (2)	20	1.742	–	0.3	89.4	10.3	S-Cs
Clay (1)	61	1.718	7.18	0.6	2.0	97.4	CH
Clay (2)	40	1.776	–	0.0	0.0	100.0	CH
Volcanic soil	88	1.393	6.16	1.7	5.8	92.5	VH2

Table 3.3 Chemical reagents mixed with the stabilized soils tested for leaching of hazardous substances.

Hazardous substance	Chemical substance	Chemical formula
Cadmium	Cadmium nitrate	$Cd(NO_3)_2.4H_2O$
Lead	Lead(II) nitrate	$Pb(NO_3)_2$
Hexavalent chromium	Potassium dichromate	$K_2Cr_2O_7$
Arsenic	Disodium hydrogenarsenate	$Na_2HAsO_4.7H_2O, KAsO_2$
Mercury	Mercuric chloride	$HgCl_2$
Selenium	Sodium selenite	Na_2SeO_4
Fluorine	Potassium fluoride	$KF.2H_2O$
Boron	Sodium metaborate	$NaBO_2.4H_2O$
Trichloroethylene	Trichloroethylene	C_2HCl_3

(Figure 3.45(b)), the amount of leaching decreases and becomes lower than the detection limit for all the stabilized soils when the cement factor is larger than 100 kg/m^3 and cured for 28 days. For the leaching of hexavalent chromium (Figure 3.45(c)), the amounts of leaching decrease only slightly with the cement factor, and the improvement effect varies depending upon the type of soil. For leaching of arsenic (Figure 3.45(d)), the amount of leaching is variable for soil type: it is decreased by the stabilization for sand (2) and clay (2). For the leaching of mercury (Figure 3.45(e)), the amounts of leaching decrease rapidly as long as the cement factor is about 100 kg/m^3, but increases with a further increase in the cement factor. This phenomenon can be seen especially in the volcanic soil. For the leaching of selenium (Figure 3.45(f)), the amounts of leaching decrease very slightly, even if the cement factor increases to 300 kg/m^3. For the leaching of fluorine and boron (Figures 3.45(g) and 3.45(h)), the amounts of leaching are variable depending on soil type: there is a decrease by stabilization for sand (1) and clay (1) but only a slight decrease for volcanic soil.

   According to the test results, the improvement effect by admixture stabilization is variable depending upon the type of soil and type of substances. A high improvement effect is achieved for cadmium and lead where the amount of leaching can be reduced lower than the soil elution criteria, as defined by the Act. For the other substances, the effect of stabilization is variable depending upon the type of soil and the amount of binder.

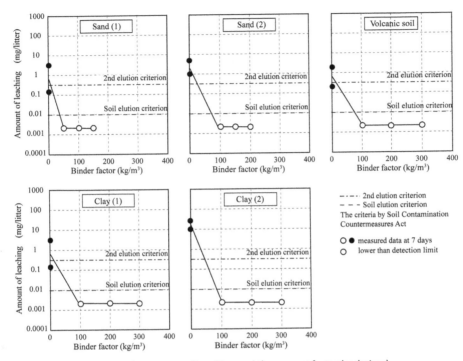

(a) Relationship between the amount of leaching and the cement factor (cadmium).

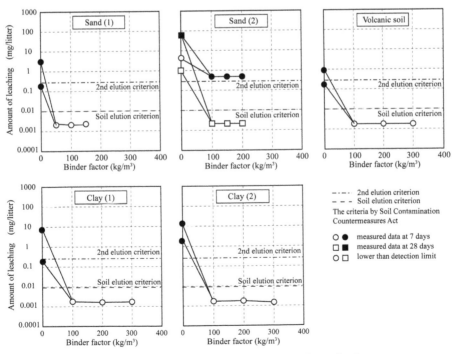

(b) Relationship between the amount of leaching and the cement factor (lead).

Figure 3.45 The effect of cement stabilization on leaching of hazardous substances from different soil types (Kaneshiro et al., 2006).

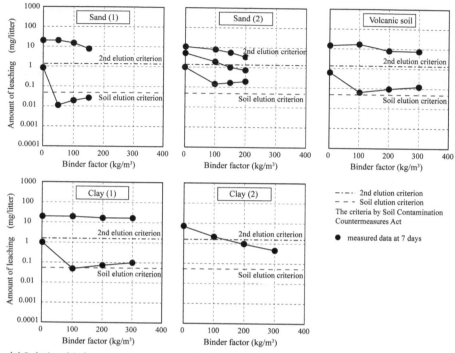

(c) Relationship between the amount of leaching and the cement factor (hexavalent chromium).

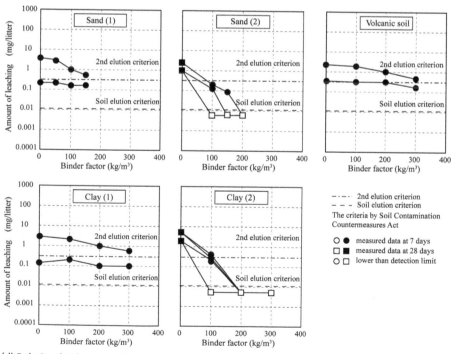

(d) Relationship between the amount of leaching and the cement factor (arsenic).

Figure 3.45 (Continued).

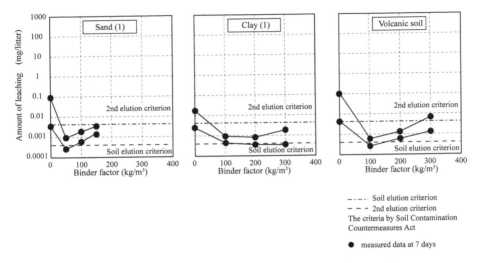

(e) Relationship between the amount of leaching and the cement factor (mercury).

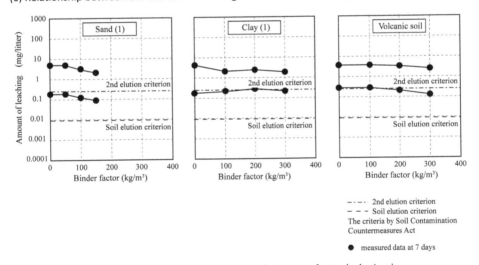

(f) Relationship between the amount of leaching and the cement factor (selenium).

Figure 3.45 (Continued).

3.4.2 Elution of hexavalent chromium (chromium VI) from stabilized soil

Figure 3.46 shows the influence of the type of binder on the elution of hexavalent chromium (chromium VI) from stabilized soils (Hosoya, 2002). In the tests, six soils including two sandy soils, two cohesive soils and two volcanic cohesive soils, were stabilized with four types of binder: ordinary Portland cement, blast furnace slag cement type B, and two cement-based special binders. The leaching tests were carried out on the stabilized soils according to the testing procedure specified by the

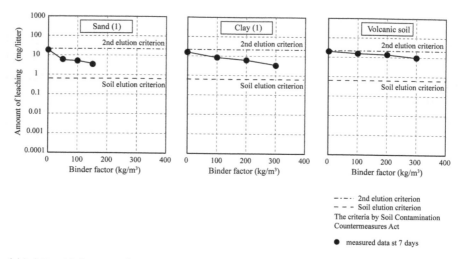

(g) Relationship between the amount of leaching and the cement factor (fluorine).

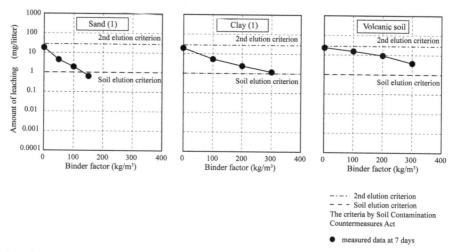

(h) Relationship between the amount of leaching and the cement factor (boron).

Figure 3.45 (Continued).

environmental quality standards (Environment Agency, 1975) and the amount of hexavalent chromium was measured by the ultrasonic extraction-diphenylcarbazide colorimetry specified by the Japanese Industrial Standard (Japan Industrial Standard, 2010). The time difference after adding sulphuric acid to diphenylcarbazide is changed: it is either 1 minute (DC1min) or 5 minutes (DC5min). The broken lines in Figure 3.46 corresponds to 0.05 mg/l, which is the soil elution criterion specified by the Japanese Ministry of Environment. The measured elution amounts of Cr(VI) are the total amount eluted from not only the original soil but also the binder. The measured value increases with the cement factor in some cases. The stabilized soils with the

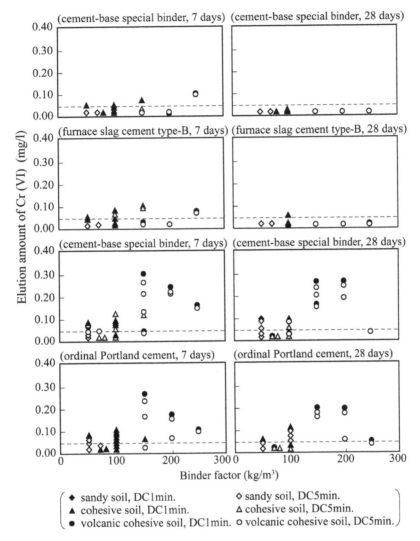

Figure 3.46 Leaching test results of hexavalent chromium from laboratory stabilized soils (Hosoya, 2002).

cement-based special binders or blast furnace slag cement type B show lower elution amounts of hexavalent chromium, Cr(VI), than those stabilized with ordinary Portland cement. As for the type of soil, the volcanic cohesive soils show the largest elution amount among the soils.

According to the accumulated test results, the leaching of hexavalent chromium is prominent in the case where the soil is volcanic soil and in an unsaturated condition, and the binder is ordinary Portland cement. The Ministry of Land, Infrastructure, Transport and Tourism, Japan, notified "the legal action on the leaching hexavalent chromium from stabilized soil" in 2000, where a laboratory test should be carried

Table 3.4 pH values of stabilized soils for different curing periods (Japan Cement Association, 2007).

Cement factor	75 kg/m^3			150 kg/m^3		
Curing period	3 days	7 days	28 days	3 days	7 days	28 days
Soil A (pH = 8.3)	12.0	11.6	11.4	12.5	12.0	11.7
Soil B (pH = 8.8)	11.7	11.3	11.2	12.0	11.7	11.6

out on the leaching of hexavalent chromium from stabilized soil to assure the amount of leaching should be lower than the criteria designated by the Soil Contamination Countermeasures Act (1975, 2005) (Table 3.1). Several types of special binder have been developed and are available on the Japanese market for mitigating the leaching of hexavalent chromium from stabilized soil.

3.4.3   Resolution of alkali from stabilized soil

When calcium hydroxide, $Ca(OH)_2$, created by the hydration of cement dissociating in water, the solution shows high alkalinity as shown in Table 3.4 (Japan Cement Association, 2007). The exposed surface of cement-stabilized soil is gradually neutralised by carbonation due to carbon dioxide in the air and the dissolution of alkali components. The alkali component dissolved from stabilized soil is not diffused widely in surrounding soil due to its buffer action.

Figure 3.47 shows the test apparatus and the measured potential hydrogen, pH, of cement-stabilized soil, the surface water (water run off the surface of stabilized soil without permeation), and the permeated water, along with the curing period (Japan Cement Association, 2007). The stabilized soil and the permeated water through stabilized soil show high pH values for three months, but the permeated water through the unstabilized soil shows neutral in pH. The surface water shows high pH values at first but gradually decreases in pH and is almost neutral after three months.

Figure 3.48 shows the pH value distributions in the cement-stabilized soil and unstabilized soil in the field, which were measured at 33 months after the stabilization (Japan Cement Association, 2007). The stabilized soil still shows a high pH value, of the order of 10 to 12, but comparatively low pH value at the shallow depth probably due to the dissolution by rainfall and surface water. In the unstabilized soil, the pH value is very high close to the boundary with the stabilized soil, but rapidly decreases with depth to a constant level at about 100 mm from the boundary.

3.4.4   Resolution of dioxin from stabilized soil

Figure 3.49 shows the relationship between the dioxin content and amount of cement in stabilized soil (Matsumura, 2007). A dredged organic soil with a natural water content of 284.4% was stabilized with cement of 100, 200 and 300 kg/m^3. At 28 days' curing, a dioxin test was carried out on the stabilized soil according to the Manual on Determination of Dioxins in Bottom Sediment, specified by Environment Agency (2008). The figure shows the amount of dioxin decreases with the cement factor.

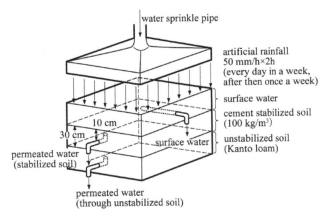

(a) Test apparatus.

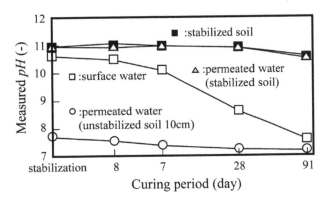

(b) Measured pH values.

Figure 3.47 Resolution of alkali from stabilized soil (Japan Cement Association, 2007).

4 PROPERTIES OF STABILIZED SOIL SUBJECTED TO DISTURBANCE/COMPACTION

4.1 Physical properties

4.1.1 Change in consistency

Figure 3.50 shows the effect of cement stabilization on the consistency of stabilized soil (Watabe & Tanaka, 2012). The Honmoku clay (w_L of 108.0% and w_P of 47.9%) was stabilized with cement factor of 100 kg/m³ and cured for 14 days. After curing, the soils were disturbed throughout and mixed again with an additional amount of cement. In the figure, the white markers show the measured values on the remixed stabilized soils, while the black markers show those on stabilized soil without remixing (see Section 4.1.3). The figure shows that the liquid limit, w_L, and the plastic limit, w_P, are increased by the first stabilization, and they increase again by remixing with

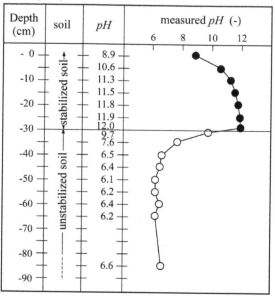

Site: parking space (Tokyo)
Curing: 2 years and 9 months

Figure 3.48 pH distribution in cement-stabilized soil and unstabilized soil at 33 months' curing (Japan Cement Association, 2007).

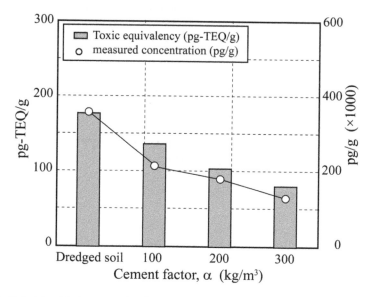

Figure 3.49 Relashionship between the dioxin content and amount of cement (Matsumura, 2007).

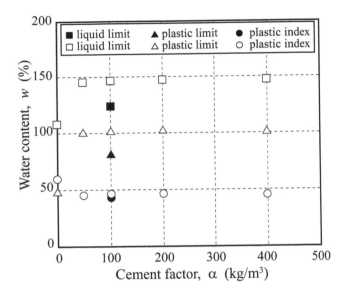

Figure 3.50 Change of consistency brought about by cement stabilization (Watabe & Tanaka, 2012).

additional cement. But they are negligibly influenced by the amount of cement in the remixing process. The plastic index is negligibly influenced by the remixing and the amount of cement.

4.2 Mechanical properties (strength characteristics)

4.2.1 Influence of soil disturbance

Figures 3.51 and 3.52 show the effect of soil disturbance on the unconfined compressive strength, q_u (Makino et al., 2014, 2015). In the tests, the kaolin clay (w_L of 77.5%, w_P of 30.3% and w_i of 120%) was stabilized with ordinary Portland cement with aw of 5 and 10%. After mixing, the stabilized soil was molded by the tapping technique according to the Japanese Geotechnical Society Standard (2009). In the case of the disturbed soil specimen, the stabilized soil mixture was stored and cured in an airtight plastic bag first to avoid any change in water content. After 3 or 7 days, the soil mixture in the bag was disturbed throughout and placed in the mold the same way as the non-disturbed soils. After the prescribed curing period, the soil samples were subjected to the unconfined compression test, as shown in the figure.

Figure 3.51 shows the stress-strain curves of non-disturbed stabilized soil and stabilized soils disturbed after 3 days and 7 days at the curing period of 28 days. The stress of the non-disturbed stabilized soil increases rapidly with the axial strain, reaching a peak strength when the axial strain is approximately 1% and then decreasing quickly after the peak strength. The stress of the disturbed stabilized soil increases gently with the axial strain, reaches a peak when the axial strain is approximately 2 to 4% and decreases gently irrespective of the cement content. The figures

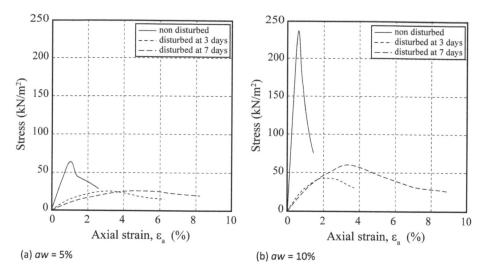

Figure 3.51 The stress-strain curves of disturbed and non-disturbed cement-stabilized soils at the curing period of 28 days (Makino et al., 2015).

reveal that the non-disturbed stabilized soil shows a brittle characteristic with quite large strength at a small axial strain and a quite small residual strength. However, the disturbed stabilized soils show a ductile characteristic with small strength and stiffness.

Figure 3.52 shows the relationship between the unconfined compressive strength and curing period (Makino et al., 2015).The unconfined compressive strength of the non-disturbed stabilized soil monotonically increases with the curing period. The unconfined compressive strength of the disturbed stabilized soils decreases considerably to 10 to 20% of that of the non-disturbed stabilized soils due to the disturbance, and it gradually increases with the curing period to approximately 25 to 40% of that of the non-disturbed stabilized soils.

A similar phenomenon on the effect of soil disturbance on the unconfined compressive strength, q_u, of cement-stabilized soil is found, in which the Nagoya Port clay (w_L of 74.4%, w_P of 33.0% and w_i of 70.0%) was stabilized with ordinary Portland cement with a cement factor, α, of 50 kg/m^3 (Ministry of Transport, The Fifth District Port Construction Bureau, 1999).

Figure 3.53 shows a similar test result to Figure 3.52, but the stabilized soil is disturbed at several curing ages up to 180 days (Japan Cement Association, 2012). The figure shows the strength can increase largely as long as the soil is disturbed at an early curing age, but the increase is quite small when it is disturbed at a late age. The strength increases and reaches an almost constant value at about four weeks in the case where the cement content is relatively small. It can be said that the strength gain is small when the soil is disturbed after the completion of cement hydration action.

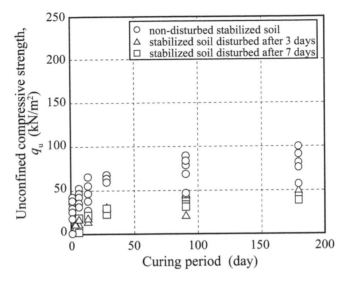

(a) *aw* = 5%.

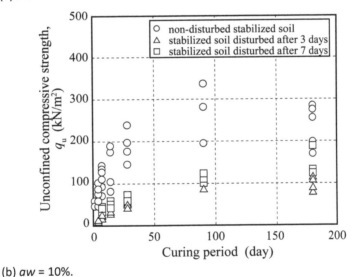

(b) *aw* = 10%.

Figure 3.52 The relationship between unconfined compressive strength and curing period of stabilized soils (Makino et al., 2015)

4.2.2 Influence of soil disturbance and compaction

Figure 3.54(a) shows the effect of the soil disturbance and compaction on the unconfined compressive strength, q_u, of cement-stabilized soil (Hino et al., 2007). In the test, the Hiro Port clay (w_L of 144.4%, w_P of 52.0%) was prepared to have an initial water content, w_i of $1.5 \times w_L$ and stabilized with either blast furnace slag cement type

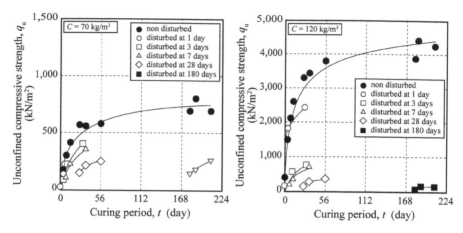

Figure 3.53 The relationship between unconfined compressive strength and the curing period of stabilized soil (Japan Cement Association, 2012).

B of 70 kg/m^3 or quicklime of 30 kg/m^3. Part of the stabilized soils were disturbed at 7 days' curing, and compacted with various compaction energies by the Japanese standard (Japanese Geotechnical Society, 2009). The figure shows that the soil strength considerably decreases to about 1/10 to 1/13 by the disturbance and increases gradually with the curing period after the compaction.

Figure 3.54(b) shows the relationship between unconfined compressive strength and compaction energy. In the case of the stabilized soil with blast furnace slag cement type B, the maximum strength can be seen at the compaction energy of 0.5 *Ec*, while the strength decreases with the compaction energy after that. In the case of the quicklime stabilization, the maximum strength can be seen at 0.25 *Ec*. They show the effect of the over-compaction phenomenon on the strength of the stabilized soils.

Figure 3.55 shows the relationship between the dry density and the compaction energy. The dry density increases with the compaction energy but soon reaches to a constant density irrespective of the type of binder and the curing period.

5   ENGINEERING PROPERTIES OF CEMENT-STABILIZED SOIL PRODUCED *IN-SITU*

5.1   Flow value of field stabilizedsoil

Figure 3.56 shows the relationship between the water content ratio and the flow value of stabilized soil (Kitazume et al., 2007). Though there is a lot of scatter in the data, the trend of data increase is roughly linear with the water content ratio.

5.2   Mixing degree of field stabilized soil

The engineering properties of stabilized soil mentioned in the previous sections were obtained mostly on laboratory stabilized soil specimens prepared with a sufficient

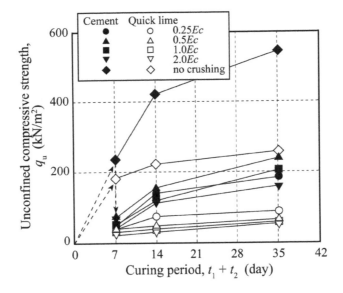

(a) Effect of curing period.

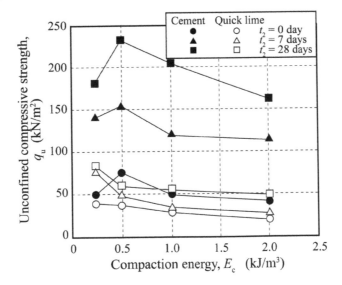

(b) Effect of compaction energy.

Figure 3.54 The effect of soil disturbance and compaction on the unconfined compressive strength of stabilized soil (Hino et al., 2007).

degree of mixing. In actual production, the original soil and binder are mixed by a pneumatic flow mixing machine and placed on land or underwater. If the mixing degree and/or the cement factor are low, the uniform mixing of original soil and binder cannot be attained in the field. If the stabilized soil is not placed by the appropriate machine

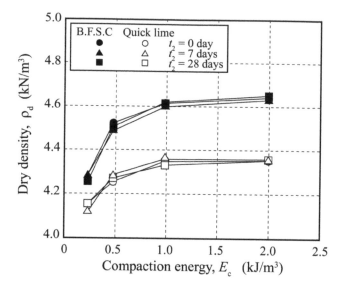

(a) Blast furnace slag cement and quicklime stabilization.

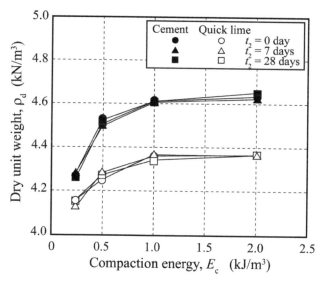

(b) Ordinary Portland cement and quicklime stabilization.

Figure 3.55 The effect of soil disturbance and compaction on the dry unit weight of stabilized soil (Hino et al., 2007).

and/or in the right manner, the uniformity and strength of the stabilized soil in the field cannot be assured to design requirements. The characteristics of field stabilized soil are, therefore, highly influenced not only by the amount of binder but also by the type of execution machine and quality control during execution. In Japan, various

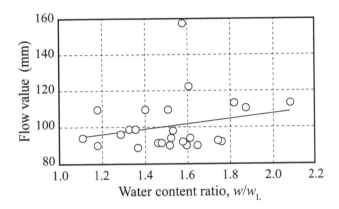

Figure 3.56 The relationship between the water content ratio and flow value of stabilized soils (Kitazume et al., 2007).

execution machines have been developed and improved, incorporating field experiences and experiments as described in Chapter 5. A careful quality control program during execution has also been developed and is practiced as routine. In this section, the characteristics of field stabilized soil produced by the Japanese machine with careful quality control are briefly introduced.

5.3   Effect of transportation distance

Figure 3.57(a) shows the relationship between the water content, w, and the unconfined compressive strength, q_u, of stabilized soil, which were sampled at four transportation distances: 43, 93, 136 and 645 m along the pipeline (Kitazume & Hayano, 2005, Hayano & Kitazume, 2005). About 25 to 30 cylindrical specimens were made from stabilized soil retrieved at each point. Meanwhile, the soil cement mixture having the same W/C as that of the retrieved plug was mixed by a mixer and poured into a cylindrical mold in a laboratory. The field and laboratory stabilized soil specimens were cured for 28 days in a laboratory, followed by the unconfined compression tests.

The two lines in the figure represent the unconfined compressive strength, q_u^{lab}, and the water content, w^{lab}, of the laboratory stabilized soil sample. The figure shows that the q_u of the stabilized soil retrieved at the distance of 43 m is significantly smaller than the q_u^{lab} and that those retrieved at the distances of 93 and 136 m are quite a lot larger than the q_u^{lab}. However, the q_u of the soil retrieved at the distance of 645 m is very close to the q_u^{lab}. The water content data from the stabilized soil is scattered quite widely in the cases of the transportation distance of 43 m and 93 m, but is less scattered over a smaller range in the case of the distance of 645 m.

Figure 3.57(b) shows the variation of average unconfined compressive strength, q_u^{ave}, of the sample along the transportation distance. In the figure, the q_u^{ave} is normalized with respect to the q_u^{lab}. The q_u^{ave} value of the stabilized soil obtained at the outlet is also plotted at the transportation distance of 1,300 m. The q_u^{ave} of the stabilized soils

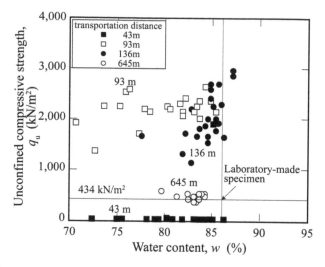

(a) Relationship between the water content and the unconfined compressive strength.

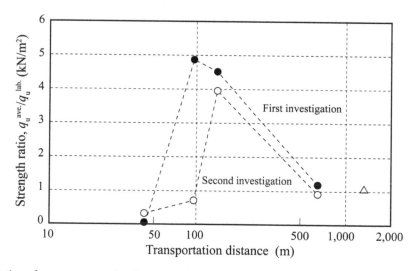

(b) Variation of average unconfined compressive strength against the transportation distance.

Figure 3.57 The effect of transportation distance on the degree of mixing of stabilized soil (Kitazume & Hayano, 2005).

retrieved at the distance of 43 m is significantly smaller than the q_u^{lab}, while that of the stabilized soil retrieved at the distance of 136 m is significantly larger. It can be assumed that the stabilized soil was sampled at the cement poor portion of the soil plug at 43 m, but at a cement-rich portion at 136 m. This fact suggests that the plug transported for a distance of less than 136 m was not thoroughly mixed so that the plug was not uniform. The samples sampled at 645 m and 1,300 m, on the other hand,

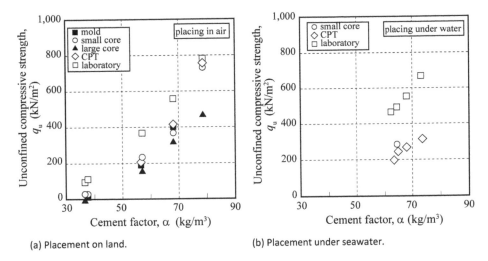

(a) Placement on land.   (b) Placement under seawater.

Figure 3.58 The unconfined compressive strength and amount of cement of field stabilized soil (Ministry of Transport, The Fifth District Port Construction Bureau, 1999).

have q_u^{ave} value very close to the q_u^{lab}. This fact suggests that a plug transported to a distance of 645 m was as well mixed as one transported to a distance of 1,300 m.

5.4 Effect of placement

5.4.1 *Effect of amount of cement on strength*

Figure 3.58 shows the relationship between the amount of cement and the unconfined compressive strength, q_u, of field stabilized soils (Ministry of Transport, The Fifth District Port Construction Bureau, 1999). The Nagoya Port clay (w_L of 74.4%, w_P of 33.0% and w_i of 127.1%) was stabilized with ordinary Portland cement with various cement factors by the pneumatic flow mixing method in the field and placed and cured either on land or under seawater. After 28 days' curing, the stabilized soils were core sampled for an unconfined compression test. The soil and cement mixture retrieved at the outlet of the pipeline and cured in a laboratory is 'mold' in the legend, and the stabilized soils core sampled in the field are 'core' and 'L core'. The specimen of 'core' is 50 mm in diameter and 100 mm in height, while the 'L core' is a large size sample, 500 mm in diameter and 1,000 mm in height. In the figure, the laboratory stabilized soil is also plotted.

In the case of placement on land, the unconfined compressive strength, q_u, increases almost linearly with the amount of cement, irrespective of the specimen type. The 'laboratory' sample shows the highest strength, but the 'L core' sample shows lowest strength, resulting in a relatively large scatter in strength. The figure also shows that at least about 40 kg/m³ of cement is necessary to obtain a certain strength gain in the field, which is very close to that of the laboratory stabilized soil as already shown in Figures 2.17 and 2.18. In the case of placement under seawater, Figure 3.59(b),

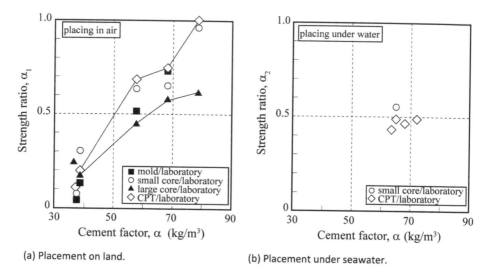

(a) Placement on land.   (b) Placement under seawater.

Figure 3.59 The unconfined compressive strength ratio and amount of cement of field stabilized soil (Ministry of Transport, The Fifth District Port Construction Bureau, 1999).

the strength of the 'core' sample is relatively small compared with the 'laboratory' sample, which could be due to the entrapped water within the stabilized soil.

Figure 3.59 shows the strength ratio of the field produced soil to the laboratory produced soil (Ministry of Transport, The Fifth District Port Construction Bureau, 1999). The strength ratios are not constant but increase with the amount of cement irrespective of field soil type, which indicates that mixing cannot be conducted well in the case of a small amount of cement. The strength ratio of the stabilized soil placed under seawater is about 20% lower than that placed on land. This phenomenon could be due to the increase of water content in stabilized soil by entrapping seawater during placement. This also emphasizes that great care should be paid during the placement of stabilized soil so as not to entrap seawater. The strength of 'L core' is lower than that of 'core' and the strength ratio of 'L core' against 'core' is about 0.7, which probably reflects the relatively large scatter in strength values. The strength ratio shown in the figure is almost the same as the previous research on the stabilized soil produced by the deep mixing method (Kitazume & Terashi, 2013).

Figure 3.60 shows the relationship between the coefficients of deviation of unconfined compressive strength, q_u, against the amount of cement (Ministry of Transport, The Fifth District Port Construction Bureau, 1999). In the figure, several field stabilized soils are plotted, together with the laboratory stabilized soil. The coefficient of deviation of the laboratory stabilized soil is relatively small, with a deviation of the order of 15%, and almost constant irrespective of the amount of cement. The field stabilized soils, on the other hand, indicate relatively large coefficients of deviation. The coefficient of variation of the small size core specimen is about 35%, irrespective of the amount of cement, which is almost of the same order as that of the deep mixing method (Kitazume & Terashi, 2013). But the 'L core' soil shows almost the same

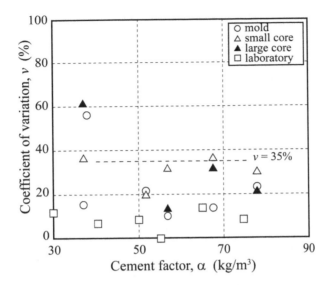

Figure 3.60 The strength deviation of field stabilized soil against the amount of cement factor (Ministry of Transport, The Fifth District Port Construction Bureau, 1999).

coefficient as the small sized specimen as long as the amount of cement exceeds about 50 kg/m³, but increases to about 60% when the amount of cement decreases to 38 kg/m³.

5.5 Heterogeneity of dredged soil

In the case of large-scale construction, dredged soil was excavated at several locations and several depths, whose properties varied very much depending on the excavation point and depth. This causes a large scatter of the stabilized soils' strength. Figures 3.61(a) and 3.61(b) provide the frequency distributions of the q_u obtained from the unconfined compression test results from the soils dredged at the areas A and B, respectively (Kitazume & Hayano, 2005, Hayano & Kitazume, 2005). The average value, $(q_u)_{ave}$, and the coefficient of variation, CV, are obtained with the normal distributions fitted to the respective data. The $(q_u)_{ave}$ of the two frequency distributions are quite similar. This is mainly because the soil properties varied along the sea depth even though the dredged soil was obtained at the same area. This fact suggests that the effect of heterogeneity of the dredged soil was that it produced varied strengths in the cement-stabilized soil. Figure 3.61(c) shows the frequency distribution of the q_u of the whole specimens of the dredged soils excavated at three areas from A to C, while the dredged soils were excavated from the five dredging areas A to E. The CV value shown in Figure 3.61(c) is similar to those shown in Figures 3.61(a) and 3.61(b). The increase of the number of dredging areas had little effect on the strength variance within the cement-stabilized soil.

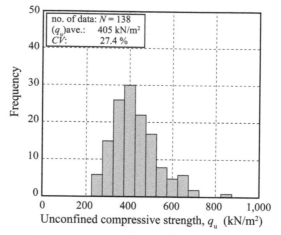

(a) Dredging area A.

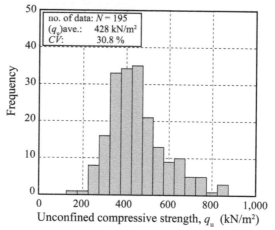

(b) Dredging area B.

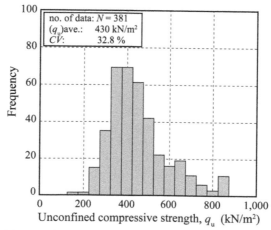

(c) Dredging area C.

Figure 3.61 The frequency distribution of q_u of field stabilized soil (Kitazume & Hayano, 2005, Hayano & Kitazume, 2005).

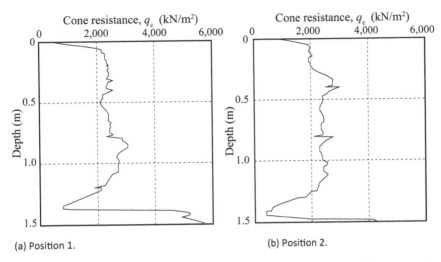

Figure 3.62 Cone penetration along depth stabilized soil ground (Ministry of Transport, The Fifth District Port Construction Bureau, 1999).

5.6 Property of stabilized ground

Figure 3.62 shows an example of the cone penetration tests on stabilized ground of about 1.5 m in thickness, where the Nagoya Port clay (w_L of 74.4% and w_P of 33.0%) was stabilized with ordinary Portland cement with various cement factors by the pneumatic flow mixing method, and placed into a pond (Ministry of Transport, The Fifth District Port Construction Bureau, 1999). Figure 3.62 shows the cone penetration test profile along the depth measured by a frictionless cone apparatus at 28 days' curing. An almost constant cone penetration test value was obtained along the whole depth, except at the most shallow layer and the bottom layer.

Figure 3.63 shows the relationship between the cone penetration test values and the unconfined compressive strength, q_u (Ministry of Transport, The Fifth District Port Construction Bureau, 1999). The figure clearly shows the cone penetration test values have a linear relation with the q_u value and its relation can be formulated as Equation 3.3. Several relationships are proposed for the coefficient between the tip resistance of the cone penetration test, q_c and q_u; $q_c = 5$ to $10 \times q_u$ for ordinary soil, and $q_c = 6.5 \times q_u$ for stabilized soil.

$$q_c = 6.5 \times q_u \tag{3.3}$$

where
q_c: CPT resistance (kN/m^2)
q_u: unconfined compressive strength (kN/m^2)

Figure 3.64 shows the plate loading test result on stabilized soil ground with five different cement factors (Ministry of Transport, The Fifth District Port Construction Bureau, 1999), where the three loading tests were carried out at around a couple

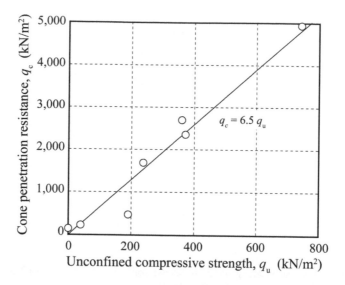

Figure 3.63 The relationship between cone penetration resistance and the unconfined compressive strength of stabilized soil ground (Ministry of Transport, The Fifth District Port Construction Bureau, 1999).

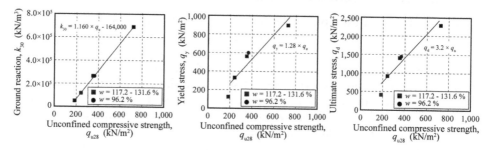

Figure 3.64 The relationship between the ground reaction and unconfined compressive strength o stabilized soil grounds with different cement factors (Ministry of Transport, The Fifth District Port Construction Bureau, 1999).

metres away from the outlet for each ground. Table 3.5 summarizes the test case and test results, where the test results in the table are the mean value of the five loading tests. The figure shows that the ground reaction factor, the yield stress and the ultimate stress of the stabilized ground, as measured in the tests, had a linear relationship with the unconfined compressive strength. According the test results, the following relations can be obtained:

$$k_{30} = 1,160 \times q_u - 164,000 \tag{3.4}$$

$$q_y = 1.28 \times q_u \tag{3.5}$$

$$q_d = 3.2 \times q_u \tag{3.6}$$

Table 3.5 Summary of plate loading tests on stabilized soil grounds (Ministry of Transport, The Fifth District Port Construction Bureau, 1999).

The test grounds	Case 1	Case 2	Case 3	Case 4	Case 5
Initial water content, w_i (%)	131.6	117.2	96.2	125.4	123.3
Cement factor (kg/m^3)	46	78	52	57	68
Unconfined compressive strength, q_{u28} (kN/m^2)	195	734	373	239	361
Yield stress, q_y (kN/m^2)	132	891	580	326	567
Ground reaction factor, K_{30} (kN/m^3)	5.37×10^4	69.0×10^4	26.3×10^4	12.7×10^4	25.9×10^4
Ultimate stress, q_d (kN/m^2)	382	2264	1415	905	1401
q_d/q_u	1.96	3.08	3.79	3.79	3.88

where

k_{30}: ground reaction coefficient (kN/m^2)

q_d: ultimate bearing capacity (kN/m^2)

q_y: yield bearing capacity (kN/m^2)

q_u: unconfined compressive strength (kN/m^2)

6 SUMMARY

The current chapter described the engineering characteristics of stabilized soil, mainly based on laboratory prepared samples. The general tendency and the correlation of various characteristics and unconfined compressive strength may apply to a variety of admixture stabilization techniques and may help design engineers understand stabilized soil.

The characteristics of *in-situ* stabilized soil discussed in Section 5, however, are only applicable to the *in-situ* soil produced by the preumatic flow mixing used in Japan. This is because the quality of *in-situ* stabilized soil depends heavily upon the mixing process and procedures. It is the responsibility of the method contractor to collect and accumulate information on the quality of *in-situ* stabilized soils produced by their own proprietary mixing system.

The knowledge compiled in the present chapter can be summarized as follows.

6.1 Properties of stabilized soil mixture before hardening

6.1.1 Physical properties

6.1.1.1 Change of consistency of the soil-binder mixture before hardening

The liquid limit decreases with the quicklime content, while the plastic limit increases. As a result, the plasticity index sharply decreases with increasing quicklime content.

6.1.2 Mechanical properties (strength characteristics)

6.1.2.1 Change in flow value

The flow value increases with the water content of a stabilized soil mixture, irrespective of the cement factor. The shear strength exponentially decreases with the flow value and a close relationship between them can be seen irrespective of the cement factor.

6.1.2.2 Change in shear strength

The shear strength of stabilized soil just after mixing is larger than that of the original soil, which is probably due to reduced fluidity due to adding cement powder. The shear strength remains a small value within about 30 minutes but increases rapidly after that due to the progress of the cement hydration effect.

6.1.2.3 Stress-strain curve

The peak compressive stress increases quickly and the axial strain at failure decreases quickly with the curing period. The brittle behavior with large peak strength, small strain at failure and small residual strength becomes more dominant with the curing period.

6.1.3 Mechanical properties (consolidation characteristics)

6.1.3.1 Void ratio -- consolidation pressure curve and consolidation yield pressure

The shapes of e-$\log p$ curves of the stabilized soil are similar to ordinary clay samples, which are characterized by a sharp bend at a consolidation yield pressure. The consolidation yield pressure is higher the longer the curing period. The relationship between the consolidation yield pressure, p_y, and the curing period, is such that the p_y increases almost linearly with the logarithm of curing period.

6.1.3.2 Coefficient of consolidation and coefficient of volume compressibility

The ratio of coefficient of consolidation of stabilized soil against original soil is 10 to 100 in a sort of over-consolidated condition, but the ratio approaches unity in a sort of normally consolidated condition. The ratio of the coefficient of volume compressibility of the stabilized soil against the original soil is 0.01 to 0.1 in a sort of over-consolidated condition, but the ratio approaches to unity in a sort of normally consolidated condition.

6.1.3.3 Coefficient of permeability

The permeability decreases with decreasing void ratio. The permeability of the stabilized soil is larger than that of the original soil. The coefficient of permeability is very much smaller than that of the original soil, and gradually decreases with the progress of cement hydration.

6.2 Properties of stabilized soil after hardening

6.2.1 Physical properties

6.2.1.1 Change in water content
The water content of the stabilized soil decreases gradually with the cement content.

6.2.1.2 Change in density
The density of the stabilized soil increases gradually with the cement content.

6.2.1.3 Change in consistency
The liquid limit and plastic limit are increased by the stabilization. As the plastic limit increases more than the liquid limit, the plasticity index is decreased by stabilization.

6.2.2 Mechanical properties (strength characteristics)

6.2.2.1 Stress-strain curve
The compressive stress increases rapidly with the axial strain to a peak strength which is followed by quick decrease. The brittle behavior with large peak strength, small strain at failure and small residual strength becomes more dominant with the curing period.

6.2.2.2 Strain at failure
The magnitude of axial strain at failure, ε_f, is of the order of 0.5 to 2.5% and considerably smaller than that of unstabilized clay. The ε_f decreases rapidly with the unconfined compressive strength, q_u.

6.2.2.3 Internal friction angle and undrained shear strength
The undrained shear strength, c_u, of the stabilized soil is larger than that of the unstabilized soil, and almost constant as long as the consolidation pressure is low. But when the consolidation pressure exceeds the consolidation yield pressure, p_y, the undrained shear strength increases with increasing consolidation pressure. The increasing ratio in the c_u of the stabilized soil is almost same as that of the unstabilized soil. The internal friction angle, ϕ', of stabilized soil is almost zero as far as the consolidation pressure is lower than the consolidation yield pressure and the same as that of the unstabilized soil when the consolidation pressure is higher than the consolidation yield pressure. The undrained shear strength influenced by the testing method.

6.2.2.4 Residual strength
The strength ratio increases with the confining pressure ratio, and is about 0.5 to 0.8 for the confining pressure ratio, $\bar{\sigma}/q_u$, exceeding about 0.1, irrespective of the type of binder.

6.2.2.5 Modulus of elasticity (Young's modulus)
The magnitude of E_{50} increases with the q_u and can be formulated as $E_{50} = 50$ to $1300 \times q_u$ for $q_u <$ about $2,000\,kN/m^2$, and $E_{50} = 350$ to $1,000 \times q_u$ for $q_u >$ about $2,000\,kN/m^2$.

6.2.2.6 Poisson's ratio

The Poisson's ratio is around 0.28 to 0.45 irrespective of the unconfined compressive strength, q_u.

6.2.2.7 Dynamic property

The initial shear modulus, G_0, increases almost linearly with the q_u and can be formulated as $G_0 = 80$ to $200 \times q_u$. The shear modulus ratio, G/G_0 is almost constant irrespective of the shear strain, while the dumping ratio, h_{eq}, increases gradually. The shear modulus ratio and dumping ratio of the stabilized soil are similar to those of the unstabilized soil.

6.2.2.8 Creep strength

The axial strain rate, $\acute{\varepsilon}$, decreases almost linearly with the time duration on the double-logarithmic graph. The decreasing phenomenon in $\acute{\varepsilon}$ is almost constant irrespective of the load intensity, q_{cr}/q_u. The stabilized soil subjected to the vertical load, q_{cr}/q_u of 0.91 exhibits creep failure at round 1 minute after loading, but the stabilized soil does not fail as long as the load intensity is lower than about $0.8 \times q_u$.

6.2.2.9 Cyclic strength

The axial compressive strain remains lower than about 0.4% in the case of $0.6 \times q_u$ even with the number of cyclic loading being 100,000. The number of cyclic loading at failure, N_f, in the logarithmic scale decreases almost linearly with the σ_{max}/q_u.

6.2.2.10 Tensile and bending strengths

The tensile strength increases almost linearly with unconfined compressive strength irrespective of the type, amount of binder and initial water content of the soil, but increments of change become smaller with increasing q_u. The tensile strength is about 0.1 to 0.6 of the unconfined compressive strength, which is influenced by the testing procedure.

6.2.2.11 Long-term strength

There are two aspects when the long-term strength of stabilized soil is concerned. One is the strength increase with time at the core portion of the stabilized soil column, which is negligibly influenced by the surrounding conditions. The other is the possible strength decrease with time in the periphery of the stabilized soil column, due to deterioration.

The long-term strength of the stabilized soil at the core increases almost linearly with the logarithm of elapsed time, irrespective of whether the stabilized soil is produced in the laboratory or in the field, the soil type or the type and amount of binder.

The long-term strength of stabilized soil at the periphery decreases with elapsed time, and the deterioration progresses gradually inward with time, especially in the case of exposure to tap water and seawater. The progress of deterioration depth in logarithmic scale is almost linear to logarithmic time, and the slopes in all the test cases are about 1/2 irrespective of the strength of specimens and the exposure conditions.

6.2.2.12   Coefficient of horizontal stress at rest

The coefficient of horizontal stress at rest, K_0, value is about 0.15.

6.2.3   Mechanical properties (consolidation characteristics)

6.2.3.1   Void ratio – consolidation pressure curve and consolidation yield pressure

The e-log p curves of the stabilized soil show a clear consolidation yield pressure and its magnitude is dependent on the cement factor. The consolidation yield pressure, p_y, has a linear relationship with the unconfined compressive strength, q_u. The ratio of p_y/q_u of the stabilized soils is 1.25.

6.2.3.2   Coefficient of volume compressibility

The coefficient of volume compressibility of the stabilized soils, m_{vs}, slightly decreases with the consolidation pressure, p. The ratio of m_{vs}/m_{vu} is 0.01 to 0.1 as long as the normalized consolidation pressure, p/p_y is around 0.1, but the m_{vs}/m_{vu} increases to unity at the p/p_y of 1.

6.2.3.3   Coefficient of consolidation

The c_{vs} value is about 10 to 100 time larger than that of unstabilized soil as long as the consolidation pressure is small, but the c_v becomes almost same as the stabilized soil when the consolidation pressure becomes large, of the order of 1,000 kN/m². The ratio of c_{vs}/c_{vu} is 10 to 100 as long as the normalized consolidation pressure, p/p_y is around 0.1, in a sort of over-consolidated condition, but the c_{vs}/c_{vu} approaches to unity at the p/p_y of 1, c_{vs}/c_{vu}, 1 when p/p_y exceeds 1.

6.2.3.4   Coefficient of permeability

The k value of the unstabilized soil is the order of around 10^{-8} to 10^{-7} cm/s and slightly decreases with consolidation pressure. The k value is larger than that of unstabilized soil when the consolidation pressure is small, but the k decreases with the consolidation pressure and becomes almost same as the unstabilized soil when the consolidation pressure becomes large, of the order of 1,000 kN/m². The coefficient of permeability is dependent upon the water content of stabilized soil and the amount of cement. The coefficient of permeability of the stabilized soil decreases with decreasing water content and with an increasing amount of cement. The coefficient of permeability of the stabilized soil decreases exponentially with increasing strength, q_u.

6.2.3.5   Secondary compression

The secondary compression coefficients, C_α, of stabilized soil decreases almost linearly with the logarithm of consolidation time, and the coefficient of the stabilized soil is about 1/3 to 1/2 of the original soil.

6.2.4   Environmental properties

6.2.4.1   Elution of contaminant

The improvement effect by admixture stabilization is variable depending upon the type of soil and type of substances. The high improvement effect is achieved for mitigating cadmium and lead from stabilized soil.

6.2.4.2   Elution of hexavalent chromium (chromium VI) from stabilized soil

The leaching phenomenon of hexavalent chromium is prominent in the case where soil is volcanic soil and in an unsaturated condition and the binder is ordinary Portland cement.

6.2.4.3   Resolution of alkali from stabilized soil

The stabilized soil still shows high pH value of the order of 10 to 12, but comparatively low pH value at a shallow depth, probably due to the dissolution of rainfall and surface water. In the unstabilized soil, the pH value is very high close to the boundary with the stabilized soil, but rapidly decreases with the depth to a constant level at about 100 mm far from the boundary.

6.2.4.4   Resolution of dioxin from stabilized soil

The amount of dioxin decreases with increasing cement factor.

6.3   Properties of stabilized soil subjected to disturbance/compaction

6.3.1   Physical properties

6.3.1.1   Change of consistency

The liquid limit, w_L, and the plastic limit, w_P, increase with the first stabilization, and they increase again by the remixing with additional cement. But they are negligibly influenced by the amount of cement in the remixing process. The plastic index is negligibly influenced by the remixing and the amount of cement.

6.3.2   Mechanical properties (strength characteristics)

6.3.2.1   Influence of soil disturbance

Non-disturbed stabilized soil shows a brittle characteristic with quite a large strength at a small axial strain and quite a small residual strength, but disturbed stabilized soils show a ductile characteristic with small strength and stiffness. The unconfined compressive strength of the non-disturbed stabilized soil monotonically increases with the curing period. The unconfined compressive strength of the disturbed stabilized soils decreases considerably, and it gradually increases with the curing period to approximately 25 to 40% of that of the non-disturbed stabilized soils.

6.3.2.2   Influence of soil disturbance and compaction

The soil strength is considerably decreased by disturbance and increases gradually with the curing period after compaction. The dry density increases with compaction energy but soon reaches a constant density irrespective of the type of binder and the curing period.

6.4   Engineering properties of field cement-stabilized soil

6.4.1   Flow value of field stabilized soil

The flow value of stabilized soil increases linearly with the water content ratio.

6.4.2 Effect of transportation distance

The property of stabilized soil is considerably influenced by the transportation distance.

6.4.3 Effect of placement

6.4.3.1 Effect of amount of cement on strength

The strength ratio of the stabilized soil placed under seawater is about 20% lower than that placed on land, which can be due to an increase of water content in stabilized soil by entrapping seawater during placement. The field stabilized soils indicate relatively large coefficients of deviation.

6.4.4 Heterogeneity of dredged soil

The effect of heterogeneity of dredged soil was large on the strength variance of the cement-stabilized soil.

6.4.5 Property of stabilized ground

The cone penetration resistance has a linear relation with the q_u value and its relation can be formulated as $q_c = 5$ to $10 \times q_u$ for ordinary soil, and $q_c = 6.5 \times q_u$ for stabilized soil. The ground reaction factor, the yield stress and the ultimate stress of the stabilized ground measured in the plate loading tests had linear relationships with the unconfined compressive strength.

REFERENCES

Coastal Development Institute of Technology (2008) *Technical Manual of Pneumatic Flow Mixing Method*, revised version. Daikousha Publishers, 188p. (in Japanese).

Enami, A., Yamada, M. & Ishizaki, H. (1993) Dynamic properties of improved sandy soils (Part 2). *Proc. of the 28th Annual Conference of the Japanese Society of Soil Mechanics and Foundation Engineering*, pp. 1065–1066 (in Japanese).

Environment Agency (1975) *Criteria for a Specific Operation of the Ground Storage Tank Outdoors using Deep Mixing Method (notification)* (in Japanese).

Environment Agency (2005) *Criteria for a Specific Operation of the Ground Storage Tank Outdoors using Deep Mixing Method (notification)* (in Japanese).

Hara, H., Suetsugu, D., Hayashi, S. & Matsuda, H. (2013) Deterioration mechanism of cement-treated soil under seawater. *Journal of Geotechnical Engineering, Japan Society of Civil Engineers*. Vol. 69, No. 4, pp. 469–479 (in Japanese).

Hayano, K. & Kitazume, M. (2005) Strength Variance within Cement Treated Soils Induced by Newly Developed Pneumatic Flow Mixing Method. *Proc. of the ACSE, Geo-Frontiers 2005*.

Hayashi, H., Nishikawa, J., Egawa, T., Terashi, M. & Ohishi, K. (2000) Long-term strength of improved column formed by Deep Mixing Method. *Proc. of the 56th Annual Conference of the Japan Society of Civil Engineers*. No. 3, pp. 378–379 (in Japanese).

Hayashi, H., Nishikawa, J., Ohishi, K. & Kitazume, M. (2003) Field observation of long-term strength of cement treated soil. *Proc. of the 3rd International Conference on Grouting and Ground Treatment*. pp. 598–609.

Hayashi, H., Ohishi, K. & Terashi, M. (2004) Possibility of strength reduction of treated soil by Ca leaching. *Proc. of the 39th Annual Conference of the Japanese Geotechnical Society*. pp. 785–786 (in Japanese).

Hino, Y., Suetsugu, D., Ikari, Y. & Ohga, T. (2007) Unconfined compressive strength, cone index and compaction properties of embankment materials by dredged-stabilized clay. *Proc. of the 42nd Annual Conference of the Japanese Geotechnical Society.* pp. 615–616 (in Japanese).

Hirade, T., M. Futaki, K. Nakano & K. Kobayashi (1995) The study on the ground improved with cement as the foundation ground for buildings, part 16. Unconfined compression test of large scale column and sampling core in several fields. *Proc. of the Annual Conference of Architectural Institute of Japan.* pp. 861–862 (in Japanese).

Hosoya, T. (2002) Leaching of hexavalent chromium from cementitious soil improvement. *Journal of Japanese Society of Materials and Science.* Vol. 51, No. 8, pp. 933–942 (in Japanese).

Ikegami, M., Ichiba, T., Ohnishi, K. & Terashi, M. (2005) Long-term property of cement treated soil 20 years after construction. *Proc. of the 16th International Conference on Geotechnical Engineering.* pp. 1199–1202.

Ikegami, M., Masuda, K., Ichiba, T., Tsuruya, H. & Ohishi, K. (2002a) Long-term durability of cement-treated marine clay after 20 years. *Proc. of the 57th Annual Conference of the Japan Society of Civil Engineers.* No. 3-61, pp. 123–124 (in Japanese).

Ikegami, M., Masuda, K., Ichiba, T., Tsuruya, H., Satoh, S., Terashi, M. & Ohishi, K. (2002b) Physical properties and strength of cement-treated marine clay after 20 years. *Proc. of the 57th Annual Conference of the Japan Society of Civil Engineers.* No. 3-61, pp. 121–122 (in Japanese).

Ikegami, M., Satoh, H., Ichiba, T., Ozawa, T., Shimura, H., Terashi, M. & Ohishi, K. (2003) Strength gain and structural change of cement treated soil. *Proc. of the 58th Annual Conference of the Japan Society of Civil Engineers.* No. 3-58, pp. 1175–1176 (in Japanese).

Japan Cement Association (2007) *Soil improvement manual using cement stabilizer (3rd edition).* Japan Cement Association, 387p. (in Japanese).

Japan Cement Association (2012) *Soil improvement manual using cement stabilizer (4th edition).* Japan Cement Association, 442p. (in Japanese).

Japan Lime Association (2009) *Technical Manual on Ground Improvement using Lime.* Japan Lime Association, 176p. (in Japanese).

Japanese Geotechnical Society (2009) *Practice for making and curing stabilized soil specimens without compaction. JGS 0821-2009.* Japanese Geotechnical Society. Vol. 1, pp. 426–434 (in Japanese).

Japanese Industrial Standard (2010) *Testing methods for industrial wastewater, JIS K 0102: 2010.* (in Japanese).

Kaneshiro, T., Moriya, M., Kondou, H. & Takahashi, S. (2006) The leaching behavior of the specific harmful substances from contaminated soil which were stabilized with all-purpose cementitious soil stabilizer. *Cement & Concrete.* No. 714, pp. 12–21 (in Japanese).

Kawasaki, T., Niina, A., Saitoh, S. & Babasaki, R. (1978) Studies on engineering characteristics of cement-base stabilized soil. *Takenaka Technical Research Report.* Vol. 19, pp. 144–165 (in Japanese).

Kitazume, M. & Hayano, K. (2005) Strength property and variance of cement treated ground by pneumatic flow mixing method. *Proc. of the 6th International Conference on Ground Improvement Techniques.* pp. 377–384.

Kitazume, M. & Takahashi, H. (2009) 27 Years' investigation on property of in situ quicklime treated clay. *Proc. of the 17th International Conference on Soil Mechanics and Geotechnical Engineering.* Vol. 3, pp. 2358–2361.

Kitazume, M. & Terashi, M. (2013) *The Deep Mixing Method.* CRC Press, Taylor & Francis Group, 410p.

Kitazume, M., Adachi, Y., Ikenoue, N. & Okubo, Y. (2007) Engineering property of treated soil by pneumatic flow mixing method (Part 3) Relation between the fluidity of treated soil and required air pressure for transport. *Proc. of the 42nd Annual Conference of the Japanese Geotechnical Society*. pp. 607–608 (in Japanese).

Kitazume, M., Jouyou, T. & Mizoguchi, M. (2006) The unconfined compressive strength of 4 years' stabilized soil strength by the pneumatic flow mixing method. *Proc. of the 61st Annual Conference of the Japan Society of Civil Engineering*, pp. 305–306 (in Japanese).

Kitazume, M., Nakamura, T., Terashi, M. & Ohishi, K. (2003) Laboratory tests on long-term strength of cement treated soil. *Proc. of the 3rd International Conference on Grouting and Ground Treatment*. Vol. 1, pp. 586–597.

Kudo, T., Seriu, M., Yoshimoto, K. & Hatakeyama, N. (1996) Strength properties of cement treated soft clay subjected to cyclic loading. *Proc. of the Symposium on Cement Treated Soils*. pp. 161–166 (in Japanese).

Makino, M., Takeyama, T. & Kitazume, M. (2014) Laboratory tests on the influence of Soil Disturbance on the Material Properties of Cement-treated Soil. *Proc. of the 9th International Symposium on Lowland Technology*. pp. 150–155.

Makino, M., Takeyama, T. & Kitazume, M. (2015) The influence of soil disturbance on material properties and micro-structure of cement-treated soil. *International Journal of Lowland Technology*.

Matsumura, M. (2007) Improvement and results of dioxin content tests of dredging soil. *Japan Society of Material Cycles and Waste Management*. Vol. 18, pp. 321–323 (in Japanese).

Ministry of Transport, The Fifth District Port Construction Bureau (1999) *Pneumatic Flow Mixing Method*. Yasuki Publishers, 157p. (in Japanese).

Miura, H., Tokunaga, S., Kitazume, M. & Hirota, N. (2004) Laboratory Permeability Tests on Cement Treated Soils. *Proc. of the International Symposium on Engineering Practice and Performance of Soft Deposits, IS-OSAKA 2004*, pp. 181–186.

Morikawa, Y., Yokoe, T. & Kitou, J. (2006) The unconfined compressive strength of 10 years' stabilized soil strength by the pneumatic flow mixing method. *Proc. of the 67th Annual Conference of the Japan Society of Civil Engineering*, pp. 451–452 (in Japanese).

Namikawa, T. & Koseki, J. (2007) Evaluation of tensile strength of cement-treated sand based on several types of laboratory tests. *Soils and Foundations*. Vol. 47, No. 4, pp. 657–674.

Niigaki, O., Fukushima, Y., Nodu, M., Yanagawa, Y. & Kasahara, Y. (2000) The property of deep mixing stabilized soil beneath highway embankment after more than 10 years. *Proc. of the 37th Annual Conference of the Japanese Geotechnical Society*. pp. 1117–1118 (in Japanese).

Niina, A., Saitoh, S., Babasaki, R., Miyata, T. & Tanaka, K. (1981) Engineering properties of improved soil obtained by stabilizing alluvial clay from various regions with cement slurry. Takenaka Technical Research Report. Vol. 25, pp. 1–21 (in Japanese).

Niina, A., Saitoh, S., Babasaki, R., Tsutsumi, I. & Kawasaki, T. (1977) Study on DMM using cement hardening agent (Part 1). *Proc. of the 12th Annual Conference of the Japanese Society of Soil Mechanics and Foundation Engineering*. pp. 1325–1328 (in Japanese).

Nishida, T., Terashi, M., Otsuki, N. & Ohishi, K. (2003) Prediction method for Ca leaching and related property change of cement treated soils. *Proc. of the 3rd International Conference on Grouting and Ground Treatment*. Vol. 1, pp. 658–669.

Onitsuka, K., Modmoltin, C., Kouno, M. & Negami, T. (2003) Effect of organic matter on lime and cement stabilized Ariake clays. *Journal of Geotechical Engineering, Japan Society of Civil Engineers*. (729/III-62), pp. 1–13.

Saitoh, S. (1988) Experimental study of engineering properties of cement improved ground by the deep mixing method. *Doctoral thesis, Nihon University*. 317p. (in Japanese).

Saitoh, S., Suzuki, Y., Nishioka, S. & Okumura, R. (1996) Required strength of cement improved ground. *Proc. of the 2nd International Conference on Ground Improvement Geosystems*. Vol. 1, pp. 557–562.

Shibuya, S., Tatsuoka, F., Teachavorasinskun, S., Kong, X. J., Abe, F., Kim, Y-S. & Park C-S. (1992) Elastic deformation properties of geomaterials. *Soils and Foundations*. Vol. 32, No. 3, pp. 26–46.

Takahashi, H. & Kitazume, M. (2004) Consolidation and Permeability Characteristics on Cement Treated Clays from Laboratory Tests. *Proc. of the International Symposium on Engineering Practice and Performance of Soft Deposits, IS-OSAKA 2004*. pp. 187–192.

Tanaka, H. & Terashi, M. (1986) Properties of Treated Soils Formed in situ by Deep Mixing Method. *Report of the Port and Harbour Research Institute*. Vol. 25, No. 2, pp. 89–119 (in Japanese).

Tatsuoka, F. & Kobayashi, A. (1983) Triaxial strength characteristics of cement treated soft clay. *Proc. of the 8th European Regional Conference on Soil Mechanics and Foundation Engineering*. Vol. 1, pp. 421–426.

Terashi, M. & Kitazume, M (1992) An Investigation of the Long-Term Strength of a Lime Treated marine Clay. *Technical Note of the Port and Harbour Research Institute*. No. 732, 14p. (in Japanese).

Terashi, M., Okumura, T. & Mitsumoto, T. (1977) Fundamental Properties of Lime-Treated Soils. *Report of the Port and Harbour Research Institute*. Vol. 16, No. 1, pp. 3–28 (in Japanese).

Terashi, M., Tanaka, H., Mitsumoto, T., Honma, S. & Ohhashi, T. (1983) Fundamental Properties of Lime and cement treated Soils (3rd Report). *Report of the Port and Harbour Research Institute*. Vol. 22, No. 1, pp. 69–96 (in Japanese).

Terashi, M., Tanaka, H., Mitsumoto, T., Niidome, Y. & Honma, S. (1980) Fundamental Properties of Lime and cement treated Soils (2nd Report). *Report of the Port and Harbour Research Institute*. Vol. 19, No. 1, pp. 33–62 (in Japanese).

The Building Center of Japan (1997) *Design and Quality Control Guideline of Improved Ground for Building*. p. 473 (in Japanese).

Udaka, K., Tsuchida, T., Imai, Y. & Tang, Y. X. (2013) Compressive characteristics of reconstituted marine clays with developed structures by adding a small amount of cement. *Japanese Geotechnical Journal*. Vol. 8, No. 3, pp. 425–439 (in Japanese).

Watabe, Y. & Tanaka, M. (2012) Study on cement proportion for recycled treated soil. *Technical Note of the Port and Airport Research Institute*. No. 1264, 8p. (in Japanese).

Watabe, Y., Tsuchida, T., Hikiyashiki, H. & Furuno, T. (2001) Mechanical and material properties of dredged soil treated with poor quality of cement. *Report of the Port and Harbour Research Institute*. Vol. 40, No. 2, pp. 3–21 (in Japanese).

Applications of the pneumatic flow mixing method

I INTRODUCTION

The pneumatic flow mixing method has many advantages, such as making beneficial use of dredged soil and subsoil possible, allowing any target stabilized soil strength to be obtained within a short period by controlling the type and amount of binder, and facilitating rapid and large-scale execution. Because of these advantages, the method has been applied to many types of construction and for many improvement purposes, including land reclamation, and backfilling behind quay walls and revetments for the purpose of reducing the earth pressure. For land reclamation applications, the binder factor and target unconfined compressive strength of stabilized soil are of the order of 50 to 70 kg/m^3 and 100 to 200 kN/m^2, respectively.

The current chapter describes some examples of the applications in Japan which will help project owners and geotechnical designers judge the applicability of the pneumatic flow mixing method to their projects.

2 IMPROVEMENT PURPOSES AND APPLICATIONS

2.1 Applications of the method

The pneumatic flow mixing method has many advantages, such as making beneficial use of dredged soil and subsoil possible, obtaining any target stabilized soil strength can be obtained within a short period by controlling the type and amount of binder, and conducting a rapid and large-scale execution. Because of these advantages, the method has been applied to many construction projects for many improvement purposes, including land reclamation; backfilling behind sea revetments and earth retaining structures; and shallow stabilization and backfill underwater, as shown in Figure 4.1 (Ministry of Transport, The Fifth District Port Construction Bureau, 1999). For these applications, ordinary Portland cement and blast furnace slag cement type B are often used as binder, where the cement factor and target unconfined compressive strength of the stabilized soil are of the order of 50 to 70 kg/m^3 and 100 to 200 kN/m^2 respectively.

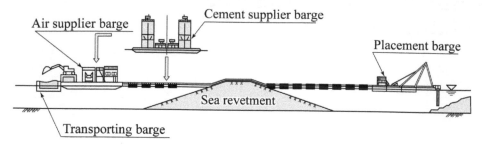

(a) Land reclamation.

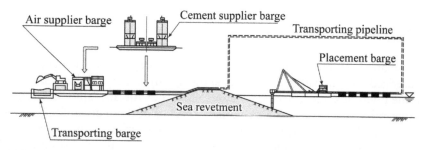

(b) Backfill behind sea revetment.

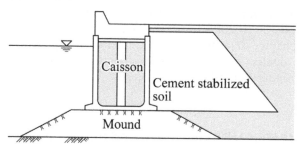

(c) Backfill behind concrete caisson.

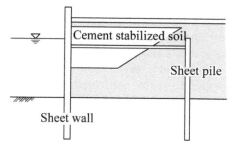

(d) Backfill behind earth retaining structure.

Figure 4.1 Examples of application of the pneumatic flow mixing method (Ministry of Transport, The Fifth District Port Construction Bureau, 1999).

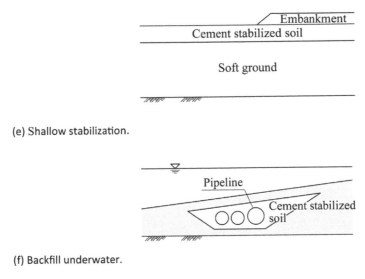

(e) Shallow stabilization.

(f) Backfill underwater.

Figure 4.1   (Continued).

3   SELECTED CASE HISTORIES OF THE METHOD IN JAPAN

As described in Chapter 1, the total number of projects of the pneumatic flow mixing method is 47 and the total volume comes to about 15.6 million m^3 from 1998 to 2015. The application of reclamation in marine areas is dominant, the proportions by number and the volume of these projects are about 68.1% and 96.6%

Among many applications of the pneumatic flow mixing method in Japan, nine case histories are selected and briefly introduced in this section: land reclamation, back fill and shallow layer construction. The locations of projects described are shown in Figure 4.2.

3.1   A field test on long-distance transport (field test)

3.1.1   Outline of project

Excavation works have been conducted every year for maintaining the sea route at Fushikitoyama Port, and the excavated soil used to be dumped in disposal sites. However, as it was anticipated that the disposal sites would be full in the near future, new disposal sites needed to be found and prepared for dredged soil. A disposal site was found on land, but it was more than 1 km away from the port. A field test was carried out at Fushikitoyama Port to investigate the capability and applicability of the pneumatic flow mixing method for long distance transportation and the strength of stabilized soil placed at the site (Uezono et al., 2000). Three types mixing techniques of the method were employed for the field test: the LMP (line mix pneumatic conveying

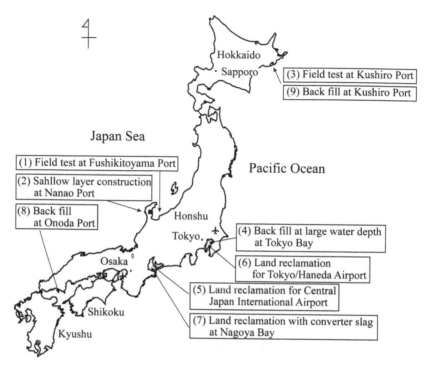

Figure 4.2 Locations of projects introduced as case histories in this chapter.

Table 4.1 Test conditions at Fushikitoyama Port (Uezono et al., 2000).

Technique	Cement factor	Total transport length	Transport length after binder injection
LMP method, slurry form binder with $W/C = 0.5$	50 kg/m³	960 m	130 m
	60 kg/m³	945 m	115 m
	80 kg/m³	930 m	100 m
Plug magic method, powder form binder	50 kg/m³	930 m	130 m
	60 kg/m³	915 m	115 m
	80 kg/m³	900 m	100 m
Snake mixing method, slurry form binder with $W/C = 1.0$	50 kg/m³	990 m	990 m
	60 kg/m³	975 m	975 m
	80 kg/m³	960 m	960 m

system) method (Sasaki, 1999), the plug magic method (Oota & Sakamoto, 2008) and the snake mixing method (Ogawa, 1999). Table 4.1 summarizes the detail of mixing techniques and execution conditions. The total of nine test cases were carried out, changing the amount of cement, as tabulated in Table 4.1.

Table 4.2 Physical properties of dredged soil and stabilized soil at Fushikitoyama Port (after Uezono et al., 2000).

	(A) LMP method			(B) Plug magic method			(C) Snake mixing method		
	slurry, $W/C = 0.5$			powder			slurry, $W/C = 1.0$		
Cement factor (kg/m^3)	50	60	80	50	60	80	50	60	80
Natural water content, w_n (%)	140.1	118.4	112.8	114.2	116.3	102.1	122.9	114.2	111.2
Flow value of natural soil	176.7	164.3	190.7	171.0	174.0	168.0	173.0	178.0	175.0
Fine particle content	85.0	77.3	69.3	81.2	78.7	71.3	76.3	70.0	70.0
Liquid limit, w_L (%)	85.3	71.9	65.1	73.5	68.0	59.3	73.9	68.4	68.2
Plastic limit, w_P (%)	31.9	28.3	24.9	30.2	28.4	26.7	33.0	26.8	25.2
Plasticity index, Ip	53	43.6	40	43	39	32	40	41	43
Ignition loss (%)	10.6	10.2	10.1	9.2	9.0	8.9	9.5	9.3	9.2
Organic matter content (%)	5.29	5.08	5.05	5.29	5.11	5.06	5.37	5.31	5.19
w_n/w_L	1.64	1.65	1.73	1.55	1.71	1.72	1.66	1.67	1.63
Water content of stabilized soil, w (%)	132.4	108.4	104.3	103.9	103.3	95.6	116.7	108.7	106.7
Flow value of stabilized soil (mm)	146.9	134.3	147.0	152.0	145.0	140.0	164.0	161.0	175.0

3.1.2 Design and stabilization work

The dredged soils used in the test, whose physical properties are summarized in Table 4.2, were excavated at Fushikitoyama Port. It was desirable to carry out the series of field tests under the same conditions, to compare the capability and applicability of the techniques to each other. However, the soil properties were varied for each mixing technique and mixing condition, unfortunately, due to differences in dredged location and depth. The field design strength at 28 curing days, q_{u28}, was determined as the relatively small value of 30 kN/m^2 by anticipating that no superstructure would be constructed on the stabilized soil ground and expected excavation in future. The laboratory mix test revealed that a cement factor of about 60 kg/m^3 should be mixed with the dredged soil to obtain the design strength. In the field test, a cement based special binder was used and its cement factor was changed 50, 60 and 80 kg/m^3 to investigate their effects on the strength of stabilized soil.

The field test was carried out in 1998, where about 100 m^3 dredged soil was stabilized by each method and mixing condition. The dredged soil was transported in the pipeline with the flow rate of 75 m^3/h and stabilized by the three techniques and placed on land by the cyclone placement method (see Chapter 5). The stabilized soil was core sampled at 7 and 28 days' curing for unconfined compression test. Table 4.3 summarizes the test results: the air pressure profile along pipeline and the unconfined compressive strength, which revealed no significant differences between each method and mixing condition. It can be concluded that all of the methods and mixing conditions have high applicability for long transportation.

The strength ratio summarized in Table 4.3(b) was defined as the strength of the field mix sample against the strength of a laboratory mix sample, q_{uf}/q_{ul}. The test

Table 4.3 Test conditions and test results (Uezono et al., 2000).

(a) Air pressure profile in pipeline.

	Air pressure (kN/m^2)		
	at inlet	at middle	at outlet
(A) LMP method			
50 kg/m³	331.2	161.7	2.9
60 kg/m³	387.1	177.4	3.9
80 kg/m³	379.3	188.2	6.9
(B) Plug magic method			
50 kg/m³	352.8	71.5	1.0
60 kg/m³	303.8	137.2	2.0
80 kg/m³	323.4	137.2	1.0
(C) Snake mixing method			
50 kg/m³	323.4	105.8	3.9
60 kg/m³	339.1	116.6	3.9
80 kg/m³	343.0	118.6	3.9

(b) Unconfined compressive strength.

	(A) LMP method slurry, $W/C = 0.5$			(B) Plug magic method powder			(C) Snake mixing method slurry, $W/C = 1.0$		
Cement factor (kg/m^3)	50	60	80	50	60	80	50	60	80
q_{ul} at 7 days (kN/m^2)	3	30	75	25	65	146	9	54	140
q_{ul} at 28 days (kN/m^2)	5	36	101	28	78	154	13	58	145
Field strength at 7 days									
q_{uf} (kN/m^2)	12	42	70	11	83	222	11	61	148
standard deviation (kN/m^2)	0.12	0.22	0.42	0.24	0.41	0.67	0.14	0.8	0.16
COV (%)	100.0	51.2	59.2	218.2	48.2	29.6	127.3	29.0	10.6
q_{uf}/q_{ul}	3.92	1.41	0.93	0.43	1.28	1.52	1.20	1.13	1.06
Field strength at 28 days									
q_{uf} (kN/m^2)	17	44	67	10	96	244	11	81	179
standard deviation (kN/m^2)	0.16	0.22	0.38	0.22	0.52	1.08	0.12	0.21	0.43
COV (%)	94.1	48.9	55.9	220.0	53.1	43.4	109.1	25.3	23.6
q_{uf}/q_{ul}	3.34	1.23	0.66	0.35	1.23	1.59	0.83	1.40	1.23

results revealed that the strength of stabilized soil increased with the cement factor, irrespective of the method. For the techniques (A) and (C), slurry types, the strength of stabilized soil was a little smaller than for technique (B). The uniformity of (A) and (C) was better than for (B), especially in (C), where the uniformity was the best among the three techniques, because the soil and cement could be mixed throughout along the long transportation distance. For (B), powder type, the strength was largest among the three techniques, but the uniformity was lower than the slurry type techniques.

Figure 4.3(a) shows the effect of water content on the unconfined compressive strength. The strength gain is quite small in the case of the cement factor of 50 kg/m³. In the case of the cement factor of 80 kg/m³, the strength decreases almost linearly

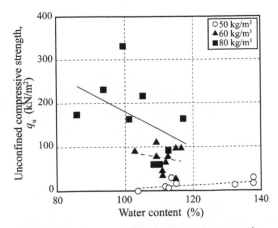

(a) Effect of water content on unconfined compressive strength.

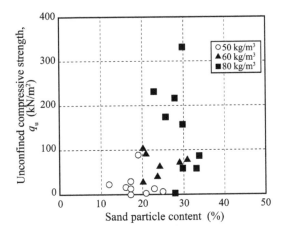

(b) Effect of sand particle content on unconfined compressive strength.

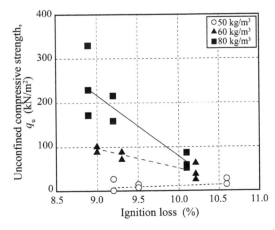

(c) Effect of ignition loss on unconfined compressive strength

Figure 4.3 Unconfined compressive strength of stabilized soil (Uezono et al., 2000).

Figure 4.4 Illustration of shipping berth at Nanao Port (Watanabe, 2005).

with the water content. Figure 4.3(b) shows the effect of sand particle content on the unconfined compressive strength. Though there is a lot of scatter in the measured data on the sand particle content, a rough phenomenon can be seen in that the strength of stabilized soil increases with the sand particle content. Figure 4.3(c) clearly shows that the strength almost linearly decreases with the ignition loss, irrespective of the cement factor.

3.2 Shallow layer construction at Nanao Port

3.2.1 Outline of project

A land reclamation project was conducted at Nanao Port, Ishikawa Prefecture, where a steel cell type berth was constructed as shown in Figure 4.4 (Watanabe, 2005). The dredged soil was backfilled behind the berth, which would be improved by the vertical drain method for accelerating the consolidation and reducing the residual settlement. As the natural water content of the dredged soil was quite high, 158.1%, it was anticipated that the stability and trafficability of a vertical drain machine could not be assured. In order to ensure stability and trafficability of the machine, stabilized soil slabs would be constructed by the pneumatic flow mixing method over the reclaimed dredged soil ground. The stabilized soil would also be applied to construct partition dikes for separating the reclaimed land into the five small ponds, 70 m × 50 m.

3.2.2 Design and stabilization work

The dredged soil excavated at Nanao Port was a very soft soil with quite a high natural water content, w_n, of 158.1%, and large liquid limit, w_L, of 177.5% and plastic limit, w_P, of 52.9%. The design strength, q_{uck}, and thickness of the stabilized soil slab were determined as 115 kN/m^2 and 1 m, respectively, by taking into account the weight of the expected vertical drain machine. The strength of the partition dike was also

Table 4.4 Soil property of original soil and mix design of stabilized soil (Watanabe, 2005).

	W/C ratio	Natural water content, w_n	Water content after adjustment	Cement factor	Volume change ratio
Partition dike	15.133	158.1%	260%	96.0 kg/m³	1.641
Slab	11.760	158.1%	370%	179.3 kg/m³	2.306

determined as the same as the slab. The mean strengths of the field stabilized soil, q_{uf}, and of laboratory stabilized soil, q_{ul}, were obtained 150.2 kN/m² and 214.6 kN/m², respectively, by assuming a coefficient of variation of 35% and a probability of 75% for the field stabilized soil, and a strength ratio, q_{uf}/q_{ul}, of 0.7.

Though the natural water content of the dredged soil was quite high, it was still smaller than the liquid limit. Some amount of water needed to be added to the dredged soil in order to obtain the sufficient fluidity of stabilized soil mixture, which was in turn anticipated to cause a decrease in the stabilized soil strength. A series of laboratory mix tests and field tests were carried out to obtain the appropriate mix condition, as summarized in Table 4.4, for assuring both the design strength and the flow value of stabilized soil.

The stabilization work was conducted in 2004, where the dredged soil was stabilized by the pipe mixing technique (Shinsha et al., 2000). In the construction of the partition dike, the stabilized soil was placed on the soft dredged soil ground by the cyclone technique first, and after one day's curing this was overlaid by the other stabilized soil placed by the direct placement technique. The stabilized soil thus placed sank into the soft dredged soil ground under its self-weight and replaced the soft ground to form the dike. In the construction of the stabilized soil slab, geotextile sheets were spread on the dredged soil ground surface first, and overlaid by the stabilized soil placed by the cyclone technique and the direct placement technique.

The stabilized soils were sampled at the outlet of pipeline and molded every day for quality control and assurance. Figure 4.5 shows the frequency distribution of the unconfined compressive strength at 28 days' curing. The mean strength of molded stabilized soils was 260 kN/m² which was larger than the laboratory strength, q_{ul}, of 214.6 kN/m². The field strength was also measured for the core samples at the partition dike, which were 120 kN/m² on average which was about 0.46 of the strength of the mold sample, probably due to the placement technique, and larger than the design strength, q_{uck}.

3.3 Field test on the strength of stabilized soil placed underwater

3.3.1 Outline of project

A field test was carried out at Kushiro West Port in 2001 to investigate the strength of stabilized soil and the mixing degree (Kobayashi et al., 2001). The dredged soil was excavated at the port and stabilized with blast furnace slug cement type B by the snake mixing method (Ogawa, 1999) and the stabilized soil was placed on the slope

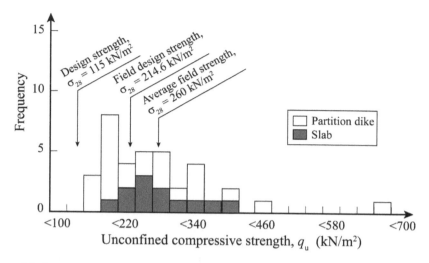

Figure 4.5 Frequency of unconfined compressive strength of stabilized soil (Watanabe, 2005).

shoulder to allow it to flow into seawater. In the tests, the snake mixer (as shown in Figure 5.7(a)) was installed at 274 m from the inlet for the cases A and at 18 m from the inlet for the cases B. The physical and mechanical properties of the stabilized soil placed underwater were investigated. The field tests revealed the high applicability of the method and the quality control in placing stabilized soil. According to the test results, the land reclamation project was conducted where more the 205,200 m³ of dredged soil (w_0 of 122.3% and w_L of 59.5%) was stabilized with blast furnace slag cement type B of the cement factor of 70 to 80 kg/m³.

3.3.2 Design and stabilization work

Figure 4.6(a) shows the relationship between the degree of mixing and the transportation distance, where the stabilized soil was collected at several points along the pipeline. The degree of mixing was defined as the strength ratio of the stabilized soil along the pipeline against one at the outlet. In the cases A, in which the snake mixer was installed at 274 m from the inlet, the degree of mixing was varied about 60% to 120% for the transportation distance was smaller than 274 m, at the location of the snake mixer, but the degree became almost constant of about 100% after the snake mixer. In the cases B in which the snake mixer was installed at 18 m from the inlet, the degree of mixing was about 50 and 80% before the snake mixer, but jumped up to about 90 to 100% after the snake mixer. The snake mixer located about 18 m from the inlet functioned very well for throughout mixing, while it had been estimated before the test that the snake mixer should be located at more than about 200 m from the inlet.

The two types of dredged soil – sandy silt and fine sand – excavated at Kushiro Port were stabilized with blast furnace slag cement type B by the snake mixing method (Ogawa, 1999) and placed on the slope shoulder to allow to flow into seawater. Figure 4.6(b) shows the strengths of stabilized soil along the distance from the outlet.

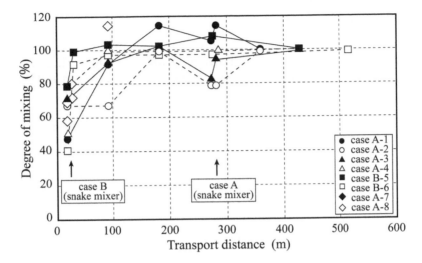

(a) Degree of mixing profile along distance from outlet.

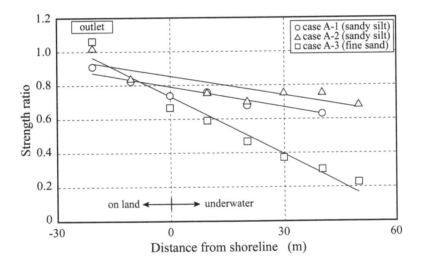

(b) Strength ratio profile along distance from outlet.

Figure 4.6 The influence of location of snake mixer on the degree of mixing and placement effect on the stabilized soil strength (Kobayashi et al., 2001).

In the cases A-1 and A-2, stabilization of sandy silt, the strength of stabilized soil decreases almost linearly with the distance, which might be due to entrapping seawater in the soil during its flow down into the seawater. The strength of stabilized fine sand in the case A-3 also decreased with distance, while the decreasing ratio was much larger than the sandy silt, in cases A-1 and A-2.

3.4   Backfill in deep water

3.4.1   Outline of project

A backfill project was conducted in Tokyo Bay, where the stabilized soil was placed behind the wharf in deep water, as shown in Figure 4.7 (Tang et al., 2000). The figure shows a section of the wharf, which consisted of a concrete caisson and rubble mound. After the backfilling, the land reclamation of 217 hectares would be carried out with dredged soil and subsoils taken in on land construction projects.

In order to prevent the dredged soil and the waste soils from leaking through the rubble mound, it is necessary to place protection inside the wharf. Use of sheets geotextile might be the first choice for the leak protection in many cases, where divers spread the sheets underwater. In the present case, however, it seemed not safe for the divers to have to spread the geotextile sheets at a depth of -40 m. Having reviewed several available methods, it was found that the dredged soft soil, after stabilization with cement, was a rational alternative for leak protection between -20 to -40 m. The design of this operation was to place the cement-stabilized soil behind the wharf, with the layer thickness greater than 1.0 m and the gradient greater than 1:3.

3.4.2   Design and stabilization work

The soft soil dredged at the site contained a certain amount of sand and its water content ranged within $85 \pm 20\%$. The unconfined compressive strength of field stabilized soil, q_{uf}, was designed to be larger than $500\,kN/m^2$, by considering the filling load afterwards. Also assuming the strength ratio of field to laboratory stabilized soil to be 0.5, the unconfined compressive strength of laboratory stabilized soil, q_{ul}, was $1,000\,kN/m^2$ and the cement factor of $90\,kg/m^3$ was determined in the laboratory mix test. The samples after curing for 28 days showed a mean unconfined compressive strength, q_{u28}, of $1,180\,kN/m^2$, which was more than double of the design strength.

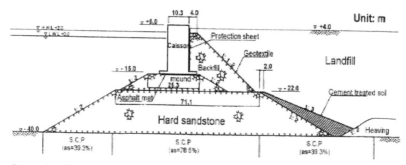

Figure 5 A section of wharf

Figure 4.7 Cross section of wharf (Tang et al., 2000).

3.5   Land reclamation for Central Japan International Airport

3.5.1   Outline of project

A man-made island was constructed for Central Japan International Airport at the north east side of Ise Bay in Nagoya, as shown in Figure 4.8 (Kitazume & Satoh, 2003, 2005; Satoh, 2003, 2004). The plane area of the man-made island was about 5.8 million m^2, and the total amount of reclamation was about 70 million m^3. It was difficult to obtain the necessary amount of soil for reclamation from the surrounding mountainous area within reasonable expense. The 8.6 million m^3 of dredged soil excavated at Nagoya Port was used as a reclamation material after stabilization by the pneumatic flow mixing method in order to reduce the amount of mountain soil needed for reclamation and to promote the recycling of dredged soil. It was the first huge scale application of the method, and a lot of laboratory and field tests were carried out to investigate the mechanical properties of stabilized soil, to develop construction machinery and procedure, and to develop quality control and assurance. These investigations were summarized and published by the Ministry of Transport in 1999 (Ministry of Transport, The Fifth District Port Construction Bureau, 1999).

3.5.2   Design and stabilization work

The designed field strength of the stabilized soil was determined as 120 kN/m^2 based on two criteria: (a) minimum *CBR* value of 3% was assured for the basement layer, and (b) no consolidation settlement was allowed. The average strength for the field stabilized soil, q_{uf}, was determined as 157 kN/m^2 by assuming the probability of 75%,

Figure 4.8 Aerial view of Central Japan International Airport.

Table 4.5 Soil property and mixing condition of expected dredged soils from Nagoya Port.

| | Property of dredged soils | | | | | | Mixing condition | | |
| | Initial water content (%) | Density (g/cm³) | Particle size distribution | | | Liquid limit (%) | Water/cement ratio W/C | Water content (%) | Cement factor (kg/m³) |
Site			gravel content (%)	sand content (%)	clay content (%)				
A	74	1.57	0.0	6.8	93.2	75.6	14.0	105	54
B	81	1.58	0.0	8.0	92.0	74.3	13.4	97	55
C	75	1.56	0.0	5.6	94.4	85.5	13.8	113	56
D	84	1.51	0.0	2.4	97.6	88.3	13.8	116	56
E	68	1.57	0.1	8.7	91.2	72.7	13.8	101	54
F	75	1.56	0.0	2.7	97.3	78.3	14.1	100	53
G	67	1.65	0.0	18.0	82.0	55.3	8.5	64	87
H	62	1.67	0.0	5.0	95.0	67.1	13.8	88	52
I	54	1.72	2.3	29.3	68.4	59.3	13.2	80	53
J	48	1.79	0.0	19.3	80.7	49.2	11.4	65	57
K	78	1.59	0.0	14.6	85.4	75.5	13.8	101	54
L	82	1.56	0.2	8.3	91.5	76.8	13.8	101	54

and the average strength of laboratory stabilized soil, q_{ul}, was determined $314\,kN/m^2$ by assuming the strength ratio, q_{uf}/q_{ul}, of 0.5.

The total of 8.6 million m³ of the dredged soil was excavated at 12 sites in Nagoya Port. The properties of the dredged soil from the different sites varies significantly, as shown in Table 4.5. For ease of quality control of the stabilized soil, a quality control system by water/cement ratio, W/C, was introduced in the project. The target flow value of the stabilized soil was determined as 95 mm, based on the previous case histories of the method (Ministry of Transport, The Fifth District Port Construction Bureau, 1999). According to these considerations and procedures, the detail mixing conditions of stabilized soil were determined for each expected dredged soil, as shown in Table 4.5.

The sea revetment was constructed in advance along all the periphery of the land reclamation area to prevent any adverse environmental impact to the surrounding seawater. Three sets of the pneumatic system were operated at the construction site to complete constructing the man-made island within about 18 months (Figure 4.9). The soft soil was mixed with blast furnace slag cement type B along the 1,500 m long pipeline and placed at the construction site by the cyclone method. The placement of stabilized soil was carried out in two stages: placement under the water level and above the water level. In the placement underwater, the stabilized soil was placed to a depth of around −5 to −1.5 m at a time. After several weeks' curing, the soil mixture was placed on top of it to the design level of +2.5 m. The stabilized soil layer was soon covered by the mountain soil to start the construction of airport facilities. The land reclamation with stabilized soil commenced in June 2001 and was completed in October 2002. The total amount of 8.6 million m³ stabilized soil was placed to construct the man-made island.

Figure 4.9 Pneumatic execution systems operating at the construction site of Central Japan International Airport.

3.5.3 Strength of the stabilized ground

Field cone tests and unconfined compression tests on core samples were performed every 25,000 m² and 40,000 m² respectively during the construction to evaluate the strength of the stabilized ground. Figure 4.10(a) shows measured unconfined compressive strength profile along the depth (Kitazume & Satoh, 2005). Though there was a lot of scatter in the measured data, the mean strength of stabilized soil was 296 kN/m², which was higher than the target value of 157 kN/m². The mean strength and the coefficient of variation of the stabilized soil placed above the water level were 364 kN/m² and 28%, respectively. Those for soil placed under the water level were 282 kN/m² and 38%, respectively, which were of smaller strength and larger variation compared to those above the water level. Figure 4.10(b) shows the water content profile of the stabilized soil along the depth. Also there was a lot of scatter in these measured data, but the mean water content was 105.3%. The water contents of the stabilized soils placed above the water level and below the water level were 96.9% and 107%, respectively, which confirms the phenomenon that seawater was entrapped during placement underwater. The entrapped seawater reduced the stabilized soil strength placed under water.

3.5.4 Long-term strength and water content of the stabilized ground

After the completion of the construction of the airport in 2002, the long-term strength and water content of the stabilized soil ground was surveyed in 2006 (4 years after the construction) (Kitazume et al., 2006) and in 2012 (10 years) (Morikawa et al., 2012). The site of the survery was at the north part of the island, as shown in Figure 4.11 (Morikawa et al., 2012). The BrN-1 and BrN-2 in Figure 4.11(a) were the survey

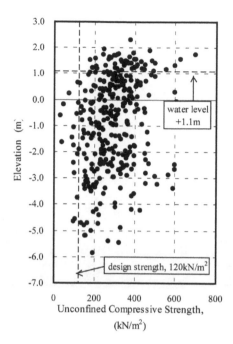

(a) Unconfined compressive strength distribution in field.

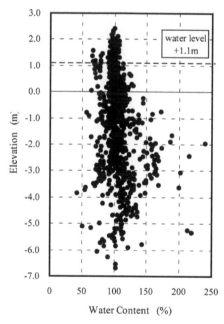

(b) Water content distribution of stabilized soil with depth.

Figure 4.10 Unconfined compressive strength and water content distribution of stabilized soil with depth (Kitazume & Satoh, 2005).

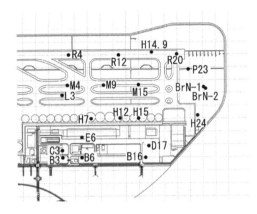

(a) Location of survey of stabilized soil in 2002, 2006 and 2012.

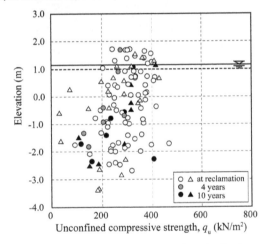

(b) Long-term unconfined compressive strength distribution along the depth.

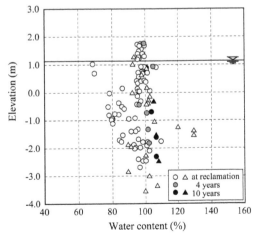

(c) Long-term water content distribution along the depth.

Figure 4.11 Long-term unconfined compressive strength and water content distribution of stabilized soil with depth (Morikawa et al., 2012).

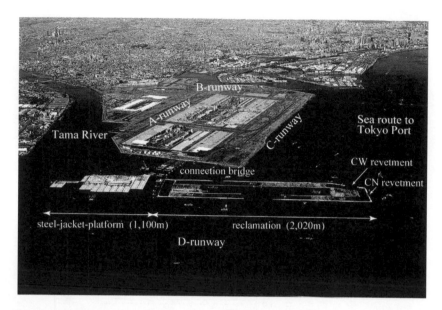

Figure 4.12 Aerial view of construction site and existing airfield on March 15th, 2009 (by courtesy of the Tokyo/Haneda International Airport Construction Office).

point in 2006 and 2012, while the others were the survey point at the completion of the construction. Figures 4.11(b) and 4.11(c) show the unconfined compressive strength, q_u, and the water content of the stabilized soils along the depth. There is a lot of scatter in the measured unconfined compressive strengths irrespective of the time of the survery. The q_u values at the reclamation varies from about 200 to 400 kN/m². The measured strengths at 4 years and 10 years are within the range of the strength at the reclamation, in which no clear decrease in the strength was found. The water content measured at 4 years and 10 years are almost a constant of about 100% as shown in Figure 4.11(c), which is within the range of those at the reclamation. Through the surveyed points in 2006 and 2012 was different from the survery at the completion of the construction, no clear change was found in the strength and the water content of the stabilized soil.

3.6 Land reclamation for Tokyo/Haneda International Airport

3.6.1 Outline of project

The construction of a fourth runway of Tokyo/Haneda International Airport was planned in 2001 and commenced in 2006 in order to cope with recent, and expected future, increases in air transportation (Figure 4.12) (Mizukami & Matsunaga, 2015). The man-made island was located between the mouth of the Tama River and the main sea route to Tokyo Port. In order to minimize any adverse influence to the water flow of the Tama River, the west part of the island was a steel-jacket-platform structure while the other was reclamation land. The east part of the island was constructed with cement-stabilized dredged soil and mountain soil. The soft soil was excavated at the

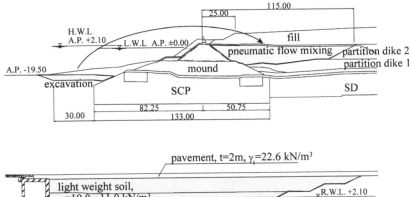

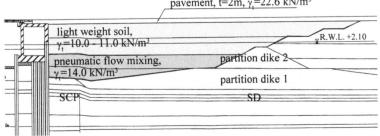

Figure 4.13 Cross section of revetment (Mizukami & Matsunaga, 2015).

front of the island and the sea route of Tokyo Port, and was stabilized with cement and backfilled behind the revetment and the steel wall revetment as shown in Figure 4.13 (Mizukami & Matsunaga, 2015).

3.6.2   Design and stabilization work

The design strength was determined as $300\,kN/m^2$ by considering that no soil failure was allowed by the overburden pressure. The coefficient of variation and the probability of the field stabilized soil were assumed as 0.35 and 75%, respectively, according to the technical manual. In order to incorporate possibility of failure of the soil due to differential settlement, the design strength was determined as the residual strength of the stabilized soil, which was evaluated at 1.20. As the strength ratio of q_{uf}/q_{ul} was assumed at 0.5, and the average strength of the laboratory stabilized soil was $942\,kN/m^2$.

The soil used in the method was excavated at three sites in the Tokyo Bay area, the properties of which are tabulated in Table 4.6. The soils had a large amount of silt and clay particles, while the fine particle content exceeded about 90%. The soils were classified as clay and silt with a high liquid limit of about 80 to 180%. Their organic matter contents were about 3.5 to 5.5%, which did not adversely affect the cement stabilization. According to the q_u and W/C relationships which were obtained in the laboratory mix tests, the mix condition for each soil were determined as shown in Table 4.6 for assuring the design strength.

In the project, about 4.8 million m^3 of soil was stabilized from autumn 2008 to the end of 2009. The stabilization work was conducted using three sets of mixing

Table 4.6 Soil properties of original soils and mixing conditions.

Soil name	Specific gravity	Target unit weight	Mixing condition	
			W/C	Cement factor
Reclamation, 1	2.662	13.01 kN/m³	9.42	90 kg/m³
Reclamation, 2	2.633	12.62 kN/m³	10.20	85 kg/m³
Reclamation, 3	2.603	12.62 kN/m³	11.17	78 kg/m³
No.1 sea route, 1	2.657	13.18 kN/m³	9.42	89 kg/m³
No.1 sea route, 2	2.630	12.76 kN/m³	7.84	109 kg/m³
No.1 sea route, 3	2.557	12.28 kN/m³	8.54	103 kg/m³
Other site	2.684	16.12 kN/m³	9.95	84 kg/m³

Figure 4.14 Placement of cement-stabilized soil in Tokyo/Haneda International Airport construction.

system. In order to minimize the water entrapment, placement was conducted by a pump system and a tremie pipe was kept within the stabilized soil at the construction site (Figure 4.14).

The stabilized soil was sampled in the field at the outlet of the placement machine and put into a plastic mold and cured in a laboratory. The stabilized soil in the field was also sampled by the rotary double core cube ($\phi = 75$ mm). The unconfined compression tests were carried out in November 2008, and Figures 4.15(a) and 4.15(b) show the unconfined compressive strength of the mold and the field stabilized soil at 91 days' curing respectively (Yamatoya et al., 2009). The mean strength of the mold stabilized soil, q_{um91}, was 1,046 kN/m², which was 111% of the target strength. The coefficient of variation of the stabilized soil and probability were 0.265 and 88.0%. The mean strength, q_{uf91}, and the coefficient of variation of the field sample were 645 kN/m² and 0.376. The strength ratio, q_{uf91}/q_{um91}, was 0.62, which was larger than the design assumed value of 0.5, and the probability of the field stabilized soil was about 85%, which was larger than the design. According to the test results on the strength ratio,

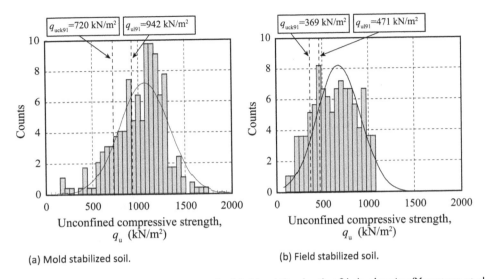

Figure 4.15 Unconfined compressive strength of field stabilized soil at 91 days' curing (Yamatoya et al., 2009).

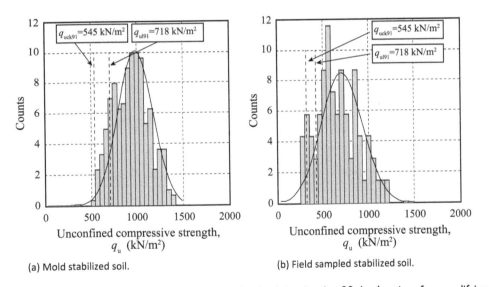

Figure 4.16 Unconfined compressive strength of field stabilized soil at 28 days' curing after modifying the mixing condition (Yamatoya et al., 2009).

the mixing condition was modified to achieve the target laboratory strength of 785 kN/m² for further stabilization work.

After modifying the mixing condition, the strength of stabilized soil was investigated again on the mold and field stabilized soils in February 2009. Figures 4.16(a) and 4.16(b) show the strength of the mold and field stabilized soils at 28 days' curing

(Yamatoya et al., 2009). The average strength, q_{um28}, the coefficient of variation, and the probability of the mold stabilized soil were $956\,kN/m^2$, 0.21, and 100%, respectively. On the field stabilized soil, the average strength, q_{uf28}, and the coefficient of variation were $672\,kN/m^2$ and 0.35, respectively, and the probability was estimated at 92.8% by incorporating the strength increase from 28 days' curing and 91 days' curing of 1.10. The strength ratio, q_{uf}/q_{um}, was 0.70, which was larger than the design, and the probability was 93%; that was also larger than the design value of 25%. Based on them, the mixing condition was modified again based on the strength ratio of 0.7.

3.7   Land reclamation using converter slag

3.7.1   Outline of project

A land reclamation project was conducted with converter steelmaking slag, so-called 'calcia improved soil', which was a mixture of dredged soil and converter steelmaking slag (Yamagoshi et al., 2013; Tanaka et al., 2012).

3.7.2   Stabilization work

The physical properties of the dredged soils excavated at Nagoya Port are shown in Table 4.7 (Yamagoshi et al., 2013). The design strength of the stabilized soil, q_{uck}, was determined as $30\,kN/m^2$. The average strengths of the field stabilized soil and laboratory stabilized soil were obtained and were calculated to be $38\,kN/m^2$ and $60\,kN/m^2$, respectively, by assuming a coefficient of variation of 0.23, probability of 75%, and a strength ratio, q_{uf}/q_{ul}, of 0.63. The laboratory mix tests were carried out changing the volume of the converter steelmaking slag and the initial water content of dredged soils. Based on the tests, the mixing ratio of the converter steelmaking slag was determined as 25% against the total volume for the stabilized soil, while the flow value of the stabilized soil was 90 to 120 mm.

Table 4.7 Physical properties of dredged soil and slag (Yamagoshi et al., 2013).

	Dredged soil (A)	Dredged soil (B)	Slag
Specific gravity, Gs	2.646	2.648	3.04
Consistency			
liquid limit, w_L (%)	100.7	58.5	–
plastic limit, w_P (%)	37.8	31.9	–
plasticity index, Ip	62.9	26.6	–
Particle size distribution			
gravel (%)	0.07	12.44	75.1
sand (%)	2.1	53.6	22.7
silt (%)	72.2	20.0	2.2
clay (%)	25.7	13.96	0
max. size (mm)	4.8	26.0	25
Wet unit weight (kN/m^3)	13.49	16.36	–
Ignition loss (%)	8.8	8.5	–
Initial water content (%)	139.0	73.3	5.2
Water content ratio, w_n/w_L	1.38	1.25	–

The dredged soil and calcia, 75:25 by volume, were mixed and transported in the pipeline, which was 800 mm in diameter and 300 m in transportation distance. The stabilized soil was placed in a storage tank by a cyclone system, and then placed at the slope shoulder by means of a concrete pump. Figure 4.17(a) shows the relationship between the flow value of stabilized soil and the gradient of air pressure in the pipeline. The flow value ranged from 82 to 156 mm due to the scatter in the initial water content. In the figure, some previous case records are plotted together. The gradient of air pressure is increased with decreasing pipeline diameter. Figure 4.17(b) shows the frequency distribution of the strength of field stabilized soil at 28 days' curing. The average strength was 114 kN/m², which exceeded the target strength of 38 kN/m².

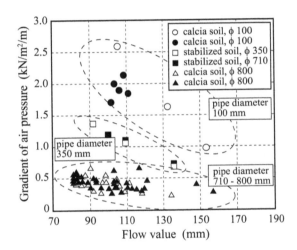

(a) Relationship between the flow value and the gradient of air pressure.

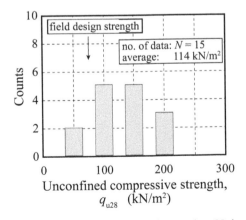

(b) Frequency distribution of the field stabilized soil strength at 28 days' curing.

Figure 4.17 Characteristics of stabilized soil in the field land reclamation test (Yamagoshi et al., 2013).

3.8   Backfill behind breakwater – for settlement reduction (field test)

3.8.1   Outline of project

Much research has been conducted to promote the beneficial use of fly ash produced in coal-fired power plants. Fly ash was used as binder to stabilize dredged soil in the following land reclamation project. The dredged soil excavated at the offshore seabed of the Onoda Power Plant was stabilized with the mixture of blast furnace slag cement type B and fly ash by the pneumatic flow mixing method (Kurumada et al., 2000). When the amount of cement is decreased to obtain a small target strength, it becomes difficult to produce a uniform stabilized soil in the field. When a low activity binder is mixed with soil, as well as cement, the volume of binder is increased and the uniformity of the stabilized soil can be improved. The fly ash was expected to function as a binder aid to obtain a relatively low strength, since it has lower hydration reactivity than cement. And fly ash was expected to increase uniformity of soil and binder mixture by the bearing effect due to the spherical shape of its particles.

3.8.2   Design and stabilization work

The properties of the dredged soil and the fly ash produced in the coal-fired power plant are shown in Table 4.8 (Kurumada et al., 2000). The design strength, q_{u28}, of the stabilized soil was determined as 100 to 250 kN/m^2 by taking into account that the stabilized soil would be excavated later for superstructure construction. A series of laboratory mix tests was carried out to obtain the mix condition, where blast furnace slag cement type B with the W/C ratio of 100% and the same amount of the fly ash as

Table 4.8   Physical properties of dredged soil and fly ash at the Onoda Power Plant (Kurumada et al., 2000).

Property	Dredged soil	Fly ash
Specific gravity, Gs	2.641	2.311
Natural water content, w_n (%)	200.68	
Grain size distribution		
sand (%)	3.4	5.6
silt (%)	42.1	81.4
clay (%)	54.5	13.0
Consistency		
liquid limit, w_L (%)	121.0	
plastic limit, w_P (%)	46.0	
plasticity index, Ip	75	
Ignition loss, L_i (%)	12.49	
Chemical composition		
SiO$_2$ (%)		45.1
Al$_2$O$_3$ (%)		23.5
Fe$_2$O$_3$ (%)		4.3
CaO (%)		13.6
MgO (%)		1.5
others (%)		12.0

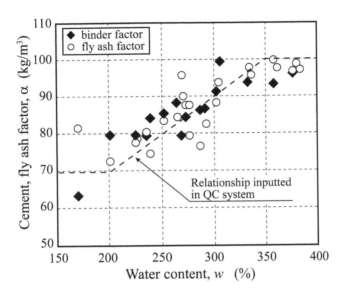

Figure 4.18 Relationship between water content and cement factor needed for the design strength.

the cement were mixed with the dredged soil. Figure 4.18 shows the test result on the relationship between the water content of the dredged soil and the amount of cement and fly ash needed to obtain the field design strength, q_{u28} (Kurumada et al., 2000). This relationship was integrated into the quality control system to control the amount of cement and fly ash for each dredged soil, according to their water content.

The stabilization work was conducted by the pneumatic flow mixing method, and controlled for each dredged soil to assure the design strength. The stabilized soil was sampled at the outlet of the placement machine and molded and cured in a laboratory, and the field stabilized soils were core sampled at 28 days' curing. Figure 4.19(a) shows the unconfined compressive strength of the samples, along with the water content of dredged soil. Through there is a lot of scatter in the strengths of the mold and core samples, they are almost constant and within the design strength range, 100 to 250 kN/m², irrespective of the water content (Kurumada et al., 2000). Figure 4.19(b) shows the frequency distribution of q_u, in which the average strength was 195.0 kN/m² and ranges from about 100 to 250 kN/m² (Kurumada et al., 2000).

3.9 Backfill behind breakwater – for settlement reduction (field test)

3.9.1 Outline of project

Stabilized soil was backfilled behind a breakwater at Kuhiro Port as shown in Figure 4.20, which was expected to increase the horizontal resistance of the break-water, promote beneficial use of dredged soil and create a growing environment for aquatic organisms (Yamauchi et al., 2011). When backfilled, the backfill soil functions to increase the horizontal stability of the breakwater which in turn can reduce

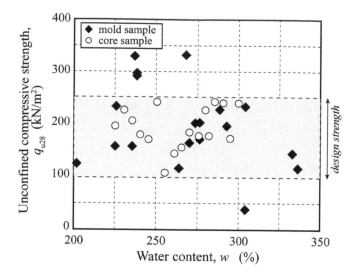

(a) Relationship between water content and cement factor.

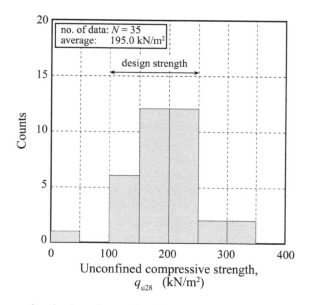

(b) Frequency distribution of unconfined compressive strength.

Figure 4.19 Characteristics of field stabilized soil (Kurumada et al., 2000).

the size of the breakwater by about 20%. When about one million m³ of dredged soil was stabilized in the project, it provided scope to prolong the disposal site on land. The sea depth was decreased to about couple of metres, which can create a growing environment for aquatic organisms.

(a) Illustration of breakwater and backfill.

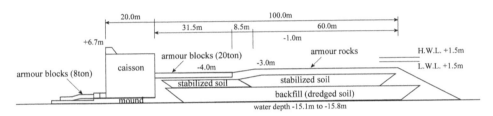

(b) Cross section of breakwater and backfill.

Figure 4.20 Schematic view of backfilling behind breakwater at Kushiro Port (Yamauchi et al., 2011).

3.9.2 Design and stabilization work

The strength of stabilized soil was designed so it had the sufficient bearing capacity for armour blocks and a mound, and so the design standard determined as q_u of 22 kN/m^2 (Ministry of Transport, 1999). The dredged soil was stabilized by the pneumatic flow mixing method and placed on land for temporary curing, and after certain curing period it was excavated and placed behind the breakwater by means of a backhoe, to minimize water pollution during the placement (Figure 4.20). As the stabilized soil was disturbed by the excavation, its strength would be smaller than that of a non-disturbed stabilized soil. In order to assure the design strength at site, the target strength of the non-disturbed stabilized soil was determined to have q_{uf} of 100 kN/m^2 by considering handing and trafficability of the soil. The laboratory strength, q_{ul}, was determined as 143 kN/m^2 by incorporating the strength ratio, q_{uf}/q_{ul}, of 0.7 according to the technical manual (Coastal Development Institute of Technology, 2001). The horizontal extent of the stabilized soil layer was designed to be about 60 m as shown in Figure 4.18 by evaluating the stability of the structure. Ordinary Portland cement was used in the project, whose cement factor was 64 kg/m^3 and 72 kg/m^3 for the dredged soil with an initial water content of 100% and 150%, respectively.

Figure 4.21 Placement of stabilized soil by a shell grab for backfill at Kushiro Port.

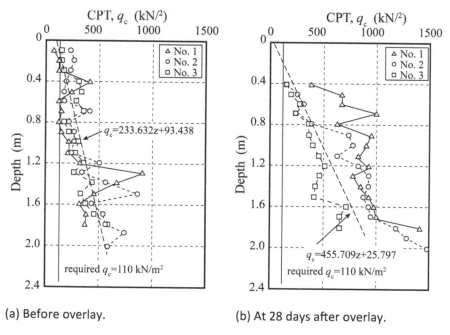

(a) Before overlay.   (b) At 28 days after overlay.

Figure 4.22 Cone penetration resistance distribution of the stabilized soil ground along depth (Honma et al., 2009).

A field test was carried out at Kushiro Port, in which a 10 m by 13 m and 2.5 m in height embankment was constructed as shown in Figure 4.21. The dredged soils excavated at Kushiro Port were organic sandy clays with organic matter content of about 3.0 to 8.7%. At 28 days' curing after placement, the stabilized soil was overlaid by the armour rocks. The cone penetration test was conducted before the overlay and at 28 days after the overlay. At 28 days after the overlay, the core samples were taken for an unconfined compression test. Figure 4.22 shows the cone penetration distribution along the depth, where the measured cone penetration resistance, q_c, was different for each testing location but increased with the depth. The measured strengths were larger than the design strength, while the target q_c was $110 \, kN/m^2$ according to the strength ratio, q_u/q_c of 0.2. The unconfined compressive strength of the core samples was $22.94 \, kN/m^2$ on average, which also satisfied the design strength.

REFERENCES

Honma, D., Ishiyama, Y. & Mori, Y. (2009) Construction of breakwater at Kushiro West area – Investigation for beneficial use of soft dredged soil to backfill behind break water. *Proc. of the 21st Research Meeting, Ministry of Land, Infrastructure, Transport and Tourism Hokkaido Regional Development Bureau.* (in Japanese).

Kitazume, M. & Satoh, T. (2003) Development of Pneumatic Flow Mixing Method and its Application to Central Japan International Airport Construction. *Ground Improvement Journal of ISSMGE.* Vol. 7, No. 3, pp. 139–148.

Kitazume, M. & Satoh, T. (2005) Quality control in Central Japan International Airport Construction. *Ground Improvement Journal of ISSMGE.* Vol. 9, No. 2, pp. 59–66.

Kitazume, M., Jouyou, T. & Mizoguchi, M. (2006) The unconfined compressive strength of 4 years' stabilized soil strength by the pneumatic flow mixing method. *Proc. of the 61st Annual Conference of the Japan Society of Civil Engineering,* pp. 305–306 (in Japanese).

Kobayashi, K., Yoshida, G. & Sato, H. (2001) Quality assurance of pneumatic flow mixing method – Snake Mixer Method -. *Proc. of the Annual Research Conference, Civil Engineering Research Institute for Cold Region, Ministry of Ministry of Land, Infrastructure, Transport and Tourism.* pp. 393–400 (in Japanese).

Kurumada, Y., Onda, T. & Saitoh, N. (2000) Construction of fly ash stabilization of high water content clay by pneumatic flow mixing method. *Proc. of the 56th Annual Conference of the Japan Society of Civil Engineers.* pp. 408–409 (in Japanese).

Ministry of Transport (1999) *Technical Standards for Port and Harbour Facilities.* The Ports and Harbours Association of Japan. pp. 525–536 (in Japanese).

Ministry of Transport, The Fifth District Port Construction Bureau (1999) *Pneumatic Flow Mixing Method.* Yasuki Publishers. 157p. (in Japanese).

Morikawa, Y., Yokoe, T. & Kitou, J. (2006) The unconfined compressive strength of 10 years' stabilized soil strength by the pneumatic flow mixing method. *Proc. of the 67th Annual Conference of the Japan Society of Civil Engineering,* pp. 451–452 (in Japanese).

Ogawa, H. (2001) Strength and mixing degree of treated soil placed underwater by pneumatic flow mixing method. *Proc. of the 56th Annual Conference of the Japan Society of Civil Engineers.* VI-213, pp. 426–427 (in Japanese).

Ogawa, H. (1999) Development of Snake Mixing Method. *Proc. of the Annual Research Meeting, Shikoku Branch, Japan Society of Civil Engineers.* VI-7, pp. 386–387 (in Japanese).

Oota, M. & Sakamoto, A. (2008) Development and application of Plug Magic method – Beneficial use of dredged soil for construction material -. *Journal of the Japanese Society of Soil Mechanics and Foundation Engineering. "Tsuchi to Kiso".* Vol. 56, No. 12, pp. 46–47 (in Japanese).

Sasaki, Y. (1999) Mud solidification air-transfer process. LMP process. *Marine Voice 21.* Vol. 205, pp. 17–19 (in Japanese).

Satoh, T. (2003) Application of pneumatic flow mixing method to Central Japan International Airport Construction. *Journal of the Japan Society of Civil Engineers.* No. 749/6-61, pp. 33–47 (in Japanese).

Satoh, T. (2004) Development and application of pneumatic flow mixing method to reclamation for offshore airport. *Technical Note of the Port and Harbour Research Institute.* No. 1076, 81p. (in Japanese).

Shinsha, H., Ikeda, S. & Matsumoto, A. (2000) Compressed air type pneumatic flow mixing method for large scale admixture stabilization of dredged clay – Pipe mixing method -. *Proc. of the 26th Annual Research Meeting, Kanto Branch, Japan Society of Civil Engineering.* pp. 1036–1037 (in Japanese).

Tanaka, Y., Yamada, K., Ookubo, Y., Shibuya, T., Nakagawa, M., Akashi, Y., Ichimura, M. & Yamagoshi, Y. (2012) Reclamation and evaluation of dredged soil with converter slag. *Journal of Geotechnical Engineering, Japan Society of Civil Engineers.* Vol. 68, No. 2, pp. 486–491 (in Japanese).

Tang, Y. X., Miyazaki, Y. & Tsuchida, T. (2000) Advanced reuses of dredging by cement treatment in practical engineering. *Prof of the International Conference of the Coastal Geotechnical Engineering in Practice.* Vol. 1, pp. 725–731.

Uezono, A., Takezawa, K., Tsukada, S. & Takahashi, K. (2000) Long distance discharging works of dredged soil by plug flow mixing method in pipe lines. *Journal of the Japan Society of Civil Engineers.* No. 651/VI-47, pp. 37–45 (in Japanese).

Watanabe, A. (2005) Challenge to soft soil – Pneumatic flow mixing method for dredged soil -. *Proc. of the Annual Research Conference, Ministry of Ministry of Land, Infrastructure, Transport and Tourism.* (in Japanese).

Yamagoshi, Y., Akashi, Y., Nakagawa, M., Kanno, H., Tanaka, Y., Tsuji, T., Imamura, T. & Shibuya, T. (2013) Reclamation of the artificial ground made of dredged soil and converter slag by using pipe mixing method. *Journal of Geotechnical Engineering, Japan Society of Civil Engineers.* Vol. 69, No. 2, pp. 952–957 (in Japanese).

Yamatoya, T., Mitarai, Y. Iba, H. & Watanabe, M. (2009) Attempt of quality control of treated soil by pneumatic flow mixing method in the construction of D runway. *Technical Meeting of Construction of Tokyo/Haneda International Airport, Ministry of Land, Infrastructure, Transport and Tourism.* (in Japanese).

Yamauchi, H., Ishiyama, Y. & Oonishi, F. (2011) Study for beneficial use of soft dredged soil – case history of back fill behind breakwater at Kushiro Port -. *Proc. of the Annual Technical Meeting, Civil Engineering Research Institute for Cold Region.* (in Japanese).

Chapter 5

Equipment, construction, and quality control and assurance

I INTRODUCTION

In this chapter, the pneumatic flow mixing equipment, construction procedure and quality control and assurance are introduced for representative pneumatic flow mixing methods in Japan. The descriptions in this chapter are based on the latest information, as of 2015. The diversified applications of the method (Chapter 4) and pursuit of cost-effectiveness have continuously promoted the improvement of existing execution systems and the development of new systems. Project owners and design engineers are encouraged to update the information periodically.

The purpose of construction by the pneumatic flow mixing method is, in many cases, to produce new ground with stabilized soil, and that the improved ground may meet the function required by geotechnical design. The responsibility for achieving the requirements are shared by owner, designer, general contractor and pneumatic flow mixing contractor, depending on the adopted contractual scheme. It is necessary for the owner and designer to have sufficient knowledge on the capability and limitation of locally available execution systems and on the experience of the local contractor, and for the contractors to understand the design intent behind the given specifications (Chapter 6).

2 EQUIPMENT

2.1 System and specifications

The system for the pneumatic flow mixing method consists of pneumatic equipment; cement plant and supplier equipment; placement equipment; and control equipment. Since quite a large amount of dredged soil at a coastal area is used to construct reclaimed land in many cases, these pieces of equipment are usually installed on barges. The system can be classified into three groups depending on the stabilization capacity: 300, 600 and 800 m^3/h. Figure 5.1 shows a group of barges for the 300 m^3/h capacity system, whose major capacities are summarized in Table 5.1 (Kitazume et al., 2000). In the figure, the dredged soil in the soil transport equipment is loaded into the hopper on the pneumatic equipment first and is transported by the help of compressed air. Binder, usually cement, is then injected into the soil on the cement supplier equipment barge and soil and binder are thoroughly mixed during transportation through the pipeline.

Figure 5.1 Group of pneumatic flow mixing barges.

Table 5.1 Major capacities of facility shown In Figure 5.1 (Kitazume et al., 2000).

Facility	Capacity
Pneumatic barge	
max. of main power	2,000 ps
max. of transporting capacity	150 m^3/h, two lines
max. of stability supplier	30 tonne/hr
Binder supplier barge	
max. capacity	300 m^3/hr
Placement barge	
diameter of cyclone	ϕ1,500 mm, two sets
Pipeline	
diameter	ϕ350 mm
length	180 to 373 m

Figure 5.2 shows the other type of barges. Binder in slurry or dry form may be added to the soil, but the slurry form is common in Japan. There are two types of the method, according to where the binder is injected, as later shown in Figure 5.4: compressor addition type and line addition type. In the former type, the binder is injected to the soil before the compressed air is injected into the pipeline. In the latter type, the binder is injected to the soil after the air injection. The soil and binder mixture is placed at the reclamation site through a cyclone on the placement barge, which functions to release the air pressure which is transporting the soil plugs. A tremie pipe is usually used to place the soil and binder mixture underwater, and thereby not entrap seawater within the soil, which can considerably weaken the stabilized soil.

(a) Soil transfer barge.

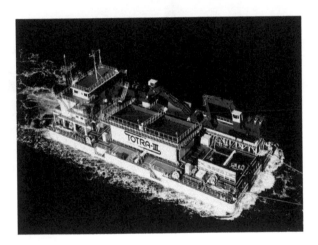

(b) Pneumatic barge.

Figure 5.2 Components of pneumatic flow mixing systems operating on floating barges.

There are several variations on the binder injection techniques and transporting techniques developed by various construction firms to improve the mixing degree of soil and binder (Hashimoto, 1999, Ikegami, 1999, Ishikawa, 2000, Iwata et al., 2000, 2006, Kobayashi et al., 2001, Mori, 2000, 2001, Ogawa, 1999, Oota & Sakamoto, 2008, Sakamoto, 1998, Sasaki, 1999, Satoh, 2001, Sato & Hayashida, 2000, Shima & Hashimoto, 1998, Sumoto, 2000). One uses an additional piece of equipment installed along the pipeline to detect the plug location accurately, and other is changing the shape or diameter of pipeline locally.

(c) Binder supplier barge.

(d) Placement barge.

Figure 5.2 Continued.

2.2 Air pressure feed system

The air pressure feed system – which has a capacity of 40 to 1,600 m³/h of compressed air supply – discharges soil into a pipeline and injects compressed air to create soil plugs, and for transporting them. It can be classified into four methods, as shown in Figure 5.3: (a) pressurized tank method, (b) mixed air feed method, (c) pressurizing pump method and (d) blower feed method.

In the pressurized tank method, soil in a pressurized tank is discharged into a pipeline by injected compressed air in the pressure tank (Figure 5.3(a)). In the mixed air feed method, soil fed continuously from an energizing device is discharged into a pneumatic pipeline by injected compressed air (Figure 5.3(b)). In the pressurizing pump method, soil fed continuously from a pressurized tank via a vertical high-speed screw is discharged into a pipeline by injected compressed air (Figure 5.3(c)).

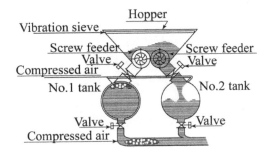

(a) Pressurized tank method.

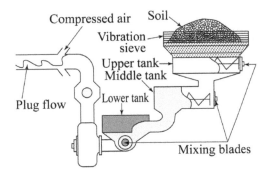

(b) Mixed air feed method.

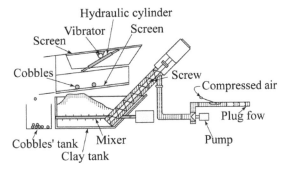

(c) Pressurizing pump method.

Figure 5.3 Air pressure feed systems for discharging soil into a pipeline.

In the blower feed method, soil fed by a rotating-type blower is discharged into a pipeline by injected compressed air (Figure 5.3(d)).

2.3   Binder supplier system

The binder supplier system injects binder to the soil. They are two types of the method according to where binder is injected, as shown in Figure 5.4: compressor addition type and line addition type. In the compressor addition type, binder is injected to the soil

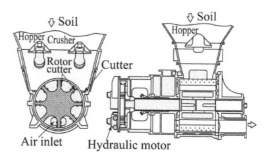

(d) Blower feed method.

Figure 5.3 Continued.

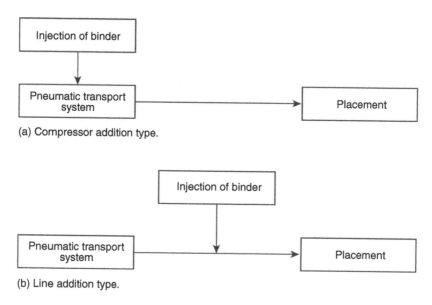

(a) Compressor addition type.

(b) Line addition type.

Figure 5.4 Schematic of binder supplier systems with different injection points.

before compressed air is injected into a pipeline (Figure 5.4(a)). A feeder with a high capacity may be necessary to feed the soil and binder mixture to the air injection point, depending on the type and property of soil, the amount of binder and the diameter of pipeline. In the line addition type, on the other hand, binder is injected to the soil in the pipeline after compressed air is injected (Figure 5.4(b)). This type has the advantage that binder may be injected at any point along the pipeline, but there needs to be a sufficiently long transportation distance from the binder injection point to the outlet to achieve a sufficient degree of mixing.

The compressor addition type can be further categorized into five systems: pipeline system, feeder system, hopper system and two sorts of tank system, as shown in Figure 5.5 (Ministry of Transport, The Fifth District Port Construction Bureau, 1999).

Snake Mixer Method

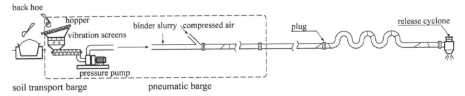

(a) Pipeline system.

Uniform soil pump method

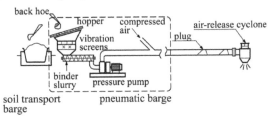

(b) Feeder system.

Rotary Wind Mixing Method

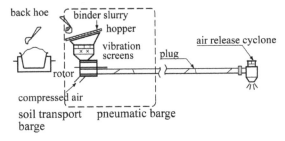

(c) Hopper system.

Figure 5.5 Compressor addition types in the binder injection system (Ministry of Transport, The Fifth District Port Construction Bureau, 1999).

In the line addition type, there are seven methods as shown in Figure 5.6. The system can also be classified according to the form of binder: dry type and wet type. In the dry type, binder in powder form is injected to the soil. In the wet type, binder is mixed with water in a slurry plant and injected to the soil, where the water to binder ratio is 60 to 100% in many cases. As shown in Figure 5.6, the wet type is applied in all the systems in Japan, expect for the plug magic line mixing method (Sakamoto, 1998, Oota & Sakamoto, 2008).

Tank & Plug Mixing Method

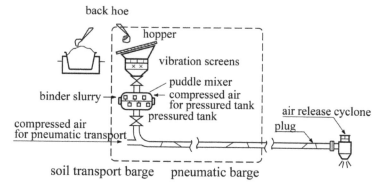

(d) Tank system.

Balance Mixing Method

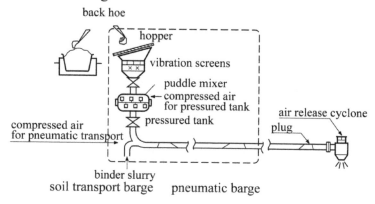

(e) Tank system.

Figure 5.5 Continued.

2.4 Pipeline

The mixture of soil and binder is transported through a pipeline, the diameter of which varies 200 mm to 800 mm depending upon the capacity of the air supplier, to the placement equipment. The minimum transportation distance necessary for complete and thorough mixing is about 100 to 200 m according to the previous case histories, as shown in Figure 4.6 (Ogawa, 2001, Kobayashi et al., 2001). There are several variations in the pipeline as shown in Figure 5.6, in which additional apparatus is installed through the pipeline with the aim of increasing mixing degree of the mixture.

Figure 5.7(a) shows the curved pipeline used in the snake mixing method, where the soil and binder mixture is agitated by the pipe's shape (Ogawa, 2001, Kobayashi

Pipe Mixing method

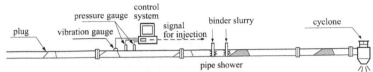

Line Mixing Pneumatic Conveying System (LMP) method

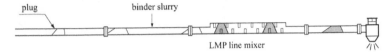

Plug Magic Line Mixing method

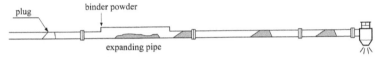

W - Tube Mixing method

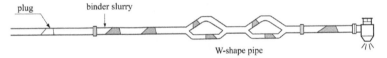

Kokusou-Diagonal Pipe Slurry Mixing (K-DPM) method

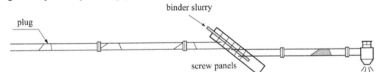

Drum Mixing method

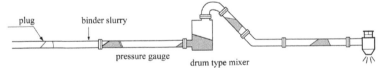

Mitsui In Line Dispenser (MILD) method

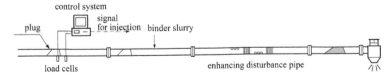

Figure 5.6 Line injection types of binder injection system (Ministry of Transport, The Fifth District Port Construction Bureau, 1999).

(a) Snake mixing method (Ogawa, 2001, Kobayashi et al., 2001).

(b) W-mixing method (Sato & Hayashida, 2000).

(c) Plug magic line mixing method (Oota & Sakamoto, 2008).

Figure 5.7 Variations of transportation pipeline.

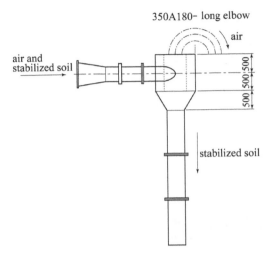

(a) Schematic profile view of air pressure release cyclone and tremie pipe.

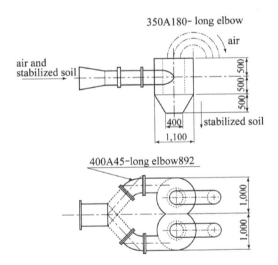

(b) Schematic profile and plane views of air pressure release cyclone.

Figure 5.8 Air pressure release cyclone and tremie pipe.

et al., 2001). Figure 5.7(b) shows the divergence in the W-mixing method, where separated soil plugs are re-merged with a time difference (Sato & Hayashida, 2000). Figure 5.7(c) shows the cement injection pipe of the plug magic line mixing method, where the diameter of the pipeline at the point of cement supply becomes locally enlarged, to 152 cm rather than the of 76 cm of the rest of the pipeline, to improve mixedness due to a temporary breaking of the plug flow (Oota & Sakamoto, 2008).

Figure 5.9 Air pressure release cyclone.

2.5   Placement equipment

The placement equipment, which can be classified into the direct placement and the cyclone methods, functions to place the stabilized soil mixture at the site. In the direct placement method, the stabilized soil mixture is discharged from the outlet vigorously. This method is very simple and does not need any machine and equipment, but the discharged soil may be segregated and can entrap water in stabilized soil in the case of placement underwater, which causes considerably weakening. In the cyclone method, which consists of a pneumatic pipe, an air pressure release cyclone and a tremie pipe. The air pressure release cyclone functions to release the air pressure and dissipate the transferring energy. Its size can be determined by the capacity of the pneumatic equipment and diameter of pipeline. Figures 5.8 and 5.9 show an example of a cyclone with a 200 m^3/h capacity.

The tremie pipe is usually used in the case of underwater placement. There are two types of tremie pipe: the fixed type and a telescoping type. In the fixed type, the length of tremie pipe is fixed. In the telescoping type, the length of tremie pipe can be adjusted according to the lift up of placed stabilized soil. The latter is preferable to prevent entrapping water within the stabilized soil, which causes considerable decrease in the stabilized soil strength.

2.6   Control equipment

Control equipment is essential in the method; during the production of stabilized soil it continuously monitors, controls and records all the quality control data, including the magnitude of air pressure and flow rate of compressed air; the unit weight and volume of the original soil; and the amount of binder (Figure 5.10). Figure 5.11 shows an example of a quality control system of the pipe mixing method. In the system, the volume and density of the original soil block are measured by means of a flow meter and a γ-ray density meter, respectively, and the measured data are transmitted to

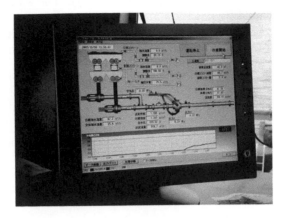

(a) PC for mixing control.

(b) Control system.

Figure 5.10 Control systems for the pneumatic flow mixing method.

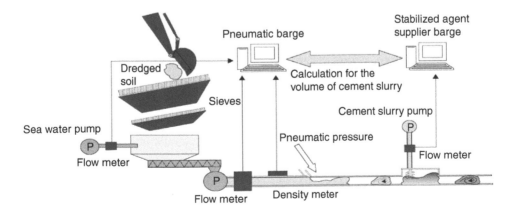

Figure 5.11 Schematic of a quality control system for the pipe mixing method.

a personal computer (PC-A) on the pneumatic barge. The water content of the original soil can be calculated by Equation 5.1. The amount of cement to be mixed with the soil block is calculated to obtain the target flow value and the target W/C value of the stabilized soil. It is transmitted to PC-B on the cement supplier barge, and the PC-B also controls the amount of cement to the soil.

3   CONSTRUCTION PROCEDURE

Regardless the construction scheme, the protocol of pneumatic flow mixing is carried out in the following steps.

(1)   Preparation of site.
(2)   Field trial test.
(3)   Construction work.

3.1   Preparation of site

In the case of a construction project on land, field preparation is carried out in accordance with the site specific conditions, which include suitable access for the plant and machinery. Before the actual operation, execution circumstances should be prepared to assure smooth execution and to prevent any adverse environmental impact.

3.2   Field trial test

It is recommended that a field trial be conducted in advance in, or adjacent to, the construction site, in order to confirm smooth execution. In the test, all the equipment for monitoring the flow rate of compressed air, the amount of binder, and the pressure and flow rate of the soil are calibrated, and it should be confirmed that the consistency and strength of the stabilized soil to meet the design requirement.

3.3   Construction work

3.3.1   Remolding and water content control

As the properties of dredged soil are usually different for any given point in the transportation barge, it is anticipated that the solid part of dredged soil may not be transported smoothly in the transportation pipeline. It can sometimes cause a blockade and then precise quality control cannot be achieved, even by the most sophisticated quality control system, where the properties of the original soil differs considerably. In order to assure the smooth transportation of dredged soil in the pipeline and increase the uniformity of character of stabilized soil, it is desirable to remold the dredged soil in the soil transportation barge to become as uniform as possible, as shown in Figure 5.12. It is desirable to measure the water content and sand particle content of the dredged soil in the transportation barge for quality control and assurance purposes.

3.3.2   Injection of binder and transportation of soil

After preliminary remolding in the transportation barge, the dredged soil is loaded onto the pneumatic barge by the backhoe, and then discharged into the pipeline.

Figure 5.12 Remolding dredged soil in the transportation barge.

Any obstacles and large stone should be removed by the vibration screen. The unit weight and volume of soil are measured by the γ-ray gauge and the flow meter, respectively. The water content of the soil is calculated by Equation 5.1 with an estimated Gs value and assuming the full saturation, $S_r = 100\%$. The volume of cement to be added, V_c, can be calculated by Equation 5.2 for the specific cement content, aw, by assuming the density of the cement, ρ_c. The soil and binder mixture is transported in the pipeline to the placement site with the help of the injected air pressure, while they are mixed throughout with the help of turbulent flow generated in the soil plugs.

$$w = \frac{1}{Gs} \cdot \frac{Gs \cdot \rho_w - \rho_t}{\rho_t - \rho_w} \cdot 100 \qquad (5.1)$$

$$V_c = \frac{aw}{100} \cdot \frac{\left(1/\rho_c + W/C/\rho_w\right) \cdot V_{soil} \cdot \rho_{soil}}{1 + w/100} \qquad (5.2)$$

where
aw : cement content (%)
Gs : specific gravity of soil particle
V_c: volume of cement slurry (m^3)
V_{soil}: volume of original soil (m^3)
w: water content of original soil (%)
W/C: water to cement ratio of slurry
ρ_c: density of cement (g/cm^3)
ρ_{soil}: density of original soil (g/cm^3)
ρ_t: density of soil particle (g/cm^3)
ρ_w: density of water (g/cm^3)

(a) Pumping type placement barge.

(b) Placement barge.

(c) Air pressure release cyclone type placement machine
(Ministry of Transport, The Fifth District Port Construction Bureau, 1999).

Figure 5.13 Placement barge and air release cyclone.

3.3.3   Placement of stabilized soil

The stabilized soil mixture, once transported through the pipeline, is placed at the site. One of the two placement methods is applied: the direct placement method or the cyclone placement method. In the direct placement method, the stabilized soil mixture is discharged from the outlet vigorously. This method is very simple and does not need any machine and equipment, but the injected soil may be segregated and can entrap water in the stabilized soil in the case of placement underwater, which causes

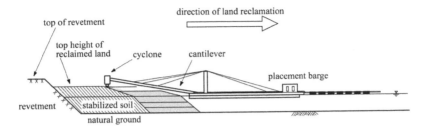

(a) Placement barge procedure.

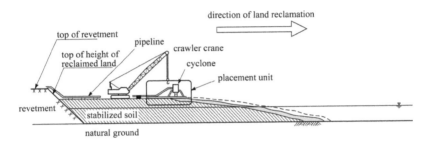

(b) Slope shoulder flow-down procedure.

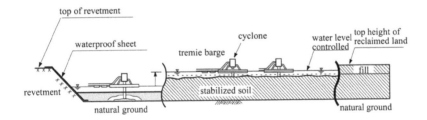

(c) Tremie barge with controlled water level procedure.

Figure 5.14 Land reclamation execution using cyclone.

a considerable decrease in the stabilized soils strength. In the cyclone method, the stabilized soil is placed gently after releasing the air pressure at a cyclone (Figure 5.13). As shown in Figure 5.18, the angle of placed stabilized soil is smaller in the direct placement method than in the cyclone placement method. It is essential in underwater placement to apply the cyclone placement method together with a tremie pipe, in order to minimize water entrapment in the soil, which causes a considerable decrease in the stabilized soil strength.

Figure 5.14 shows examples of placement procedure for land reclamation: placement barge procedure, slope shoulder flow-down procedure, and tremie placement with water level control procedure. In the placement barge procedure, Figure 5.14(a),

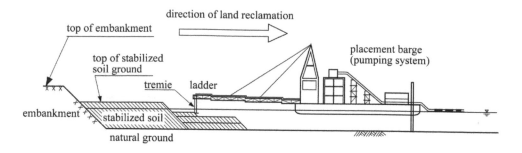

(a) Placement barge procedure.

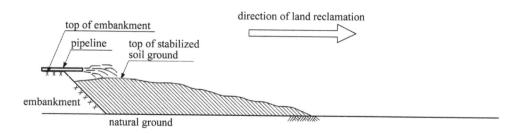

(b) Direct placement procedure.

Figure 5.15 Backfilling behind a sea revetment.

stabilized soil is placed layer by layer to the design height using a placement barge with a sufficient length of cantilever to meet a certain time interval which allows the stabilized soil to gain strength. The placement barge moves back with the progress of placement. In the slope shoulder flow-down procedure, Figure 5.14(b), stabilized soil is placed continuously to the design height with a cyclone installed on the stabilized soil ground to allow the stabilized soil to flow down the slope shoulder of a sea revetment. The placement system with a cyclone is moved forward with the progress of placement. In the tremie placement with water level control procedure, Figure 5.14(c), stabilized soil is placed layer by layer to the design height by a tremie barge with a means of controlling the water level. The tremie barge moves to place stabilized soil in flat layers.

Figure 5.15 shows several execution procedures for backfilling behind embankments. In the placement barge procedure, Figure 5.15(a), stabilized soil is placed layer by layer to the design height by a placement barge with a sufficient length of cantilever. This procedure is quite similar to the placement barge procedure as shown in Figure 5.14(a), except the stabilized soil is placed through a tremie pipe to place it underwater. In the direct placement procedure, Figure 5.15(b), the stabilized soil is discharged from the top of the sea revetment with the help of the transported air pressure. The stabilized soil discharged in this way is spread and extended in the reclamation site.

4 QUALITY CONTROL

4.1 Quality control before production

4.1.1 Soil property

The ignition loss and pH value of original soil are one of the critical parameters to investigate the applicability of the cement stabilization. In a case where the soil contains a large amount of water or organic material, ordinary Portland cement and blast furnace slag cement type B may not function well and a special binder may be necessary. The physical and chemical properties of soils should be investigated in advance. The specific gravity of soil is necessary to calculate the weight and volume of binder, as shown in Equation 5.1.

A series of laboratory mix tests should be carried out on the soils to obtain the mixing conditions, since variations in the properties and water content of soil very much depend on excavation location and depth. It is desirable to obtain the mixing condition for the soils at various locations and depths to achieve the design flow value and strength of stabilized soil. According to the accumulated case histories, the water-to-binder ratio of stabilized soil, W/C, is one of the essential parameters for the mixing design, where the unconfined compressive strength is inversely proportional to W/C, as shown in Figure 2.16.

As the amount of leachate of hexavalent chromium is regulated for cement stabilization in Japan, a leaching test needs to be carried out on the stabilized soils by the specified testing procedure (Environment Agency, 2005) and measured by the specified ultrasonic extraction, which is diphenylcarbazide colorimetry. If the amount of leachate of hexavalent chromium exceeds the regulation, the special binder should be used instead of ordinary Portland cement or blast furnace slag cement type B.

4.1.2 Pipeline length

As the soil and binder are expected to mix during transportation in the pipeline, a sufficient length of pipeline is necessary for achieving throughout mixing. As shown in Figure 4.6, the minimum transportation distance of pipeline to achieve this is 100 to 200 m. In the case where the placement site is close to the mixing plant, the pipeline should be detoured to assure the minimum distance.

4.2 Quality control during execution

4.2.1 Material control

The type, delivery date, volume and other necessary data of the supplied binder should be recorded when received. For every soil transportation barge, the physical properties of its dredged soils are evaluated, which includes the water content, bulk density and volume of the dredged soil. In the case of the wet type of binder, the water-to-binder ratio of binder slurry should be controlled to secure the prescribed value.

4.2.2 Transportation control

In order to secure the prescribed volume and properties of the stabilized soil in the field, the volume of soil transferred, the volume of air injected, and the pressure

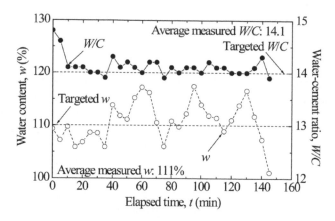

Figure 5.16 Water content and water-cement ratio in the quality control during the execution (Hayano & Kitazume, 2005).

in the pipeline should be monitored continuously and controlled precisely. Figure 5.16 shows an example of quality control data during the execution of the Central Japan International Airport construction project (Hayano & Kitazume, 2005, Kitazume & Hayano, 2007). The unit weight and volume of a soil block at the discharge point were measured by γ-ray meter and flow meter, respectively. The water content of the soil block was calculated by means of the measured unit weight. The amount of water and cement to be injected were calculated based on the quantities required to assure the target water content and W/C ratio. The water content of the soil still varied within a range of about 100 to 118%, while the target water content was 110%. The W/C ratio, on the other hand, could be well controlled to the target W/C ratio of 14. All the monitoring data – such as the excavation location and depth of the dredged soil, the amount of water supply, the amount of binder (binder slurry), and the flow rate of stabilized soil – should be recorded, together with the location of placement of stabilized soil.

4.2.3 Placement control

Based on a placement plan (reclamation plan), the placement of stabilized soil is carried out by controlling its volume and placement location. The volume, location and height of placed stabilized soil are measured and controlled. As shown in Figure 5.17, the level of the stabilized soil ground is usually measured by a movable buggy equipped with a GPS system for measuring placement on land, and by a narrow multi-beam sonar in the case of underwater placement. The measured data are transmitted to the control room to draw a contour map of the site. During the placement, the environmental issues should be monitored and controlled to prevent any adverse influence to the surrounding area. In the case of placement underwater, the placement technique influences the strength of stabilized soil very much as shown in Figures 3.58 and 3.59.

(a) Movable buggy equipped with a GPS system.

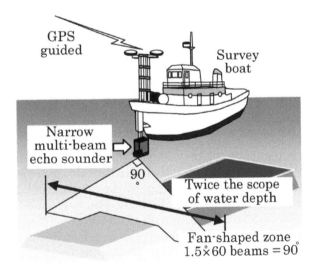

(b) Narrow multi-beam echo sounder.

Figure 5.17 Placement control in on land and marine constructions.

4.2.3.1 Slope angle

The shape of placed stabilized soil should be controlled during placement by an appropriate method. Figure 5.18 shows the relationship between the slope angle of stabilized soil placed on land and the amount of cement (Ministry of Transport, The Fifth District Port Construction Bureau, 1999). Marine clays with various initial water

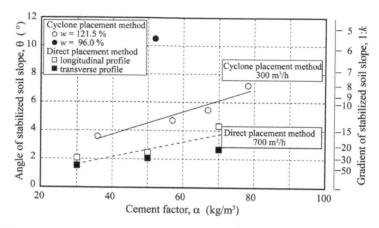

Figure 5.18 The relationship between the slope angle of stabilized soil placed on land and the amount of cement (Ministry of Transport, The Fifth District Port Construction Bureau, 1999).

Table 5.2 Execution conditions of placement methods (Ministry of Transport, The Fifth District Port Construction Bureau, 1999).

	Cyclone placement	Direct placement
Transportation distance (m)	180	315
Pipeline diameter (mm)	350	560
Total amount of stabilized soil placed (m³)	2,400	4,500
Stabilization rate (m³/h)	107 to 320	631 to 757

contents were stabilized with several amounts of blast furnace slag cement type B. The stabilized soils were placed by either cyclone or direct placement method. The execution conditions of the two methods are summarized in Table 5.2. The figure shows that the slope angle of stabilized soil is influenced by the mixing condition and placement method, and increases with the amount of cement used in its stabilization.

Figure 5.19 shows similar test results to those in Figure 5.18, in which the relationship between the slope angle of stabilized soil placed under seawater and the amount of cement used, where the execution conditions are summarized in Table 5.3. The shoot in the table is a placement techniques where the stabilized soil after the air released at the cyclone is flowed down on a shoot to placed underwater.

4.2.3.2 Strength

Figure 5.20 shows the laboratory and field tests on the effect of underwater placement on the strength of cement-stabilized soil (Tang et al., 2002). Figure 5.20(a) shows the strength ratio of the soils placed on land and underwater which was obtained in a laboratory model test. In the test, the Tokuyama clay (w_L of 77%) was mixed with water to the initial water content of 118 to 300% and slag cement of 80 to 160 kg/m³, then squeezed into the molds in air and underwater. The figure shows that the unconfined compressive strength of the underwater specimen is 92% of that

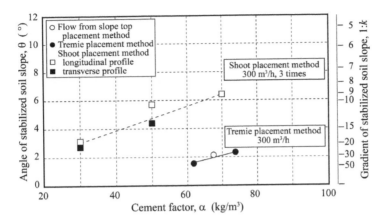

Figure 5.19 The relationship between the slope angle of stabilized soil placed under seawater and the amount of cement (Ministry of Transport, The Fifth District Port Construction Bureau, 1999).

Table 5.3 Execution conditions of placement methods (Ministry of Transport, The Fifth District Port Construction Bureau, 1999).

	Cyclone	Cyclone & tremie	Shoot
Transportation distance (m)	200	230	225
Pipeline diameter (mm)	350	350	560
Total amount of stabilized soil placed (m³)	1,195	1,800	15,200
Stabilization speed (m³/h)	298	310 to 320	548 to 683

placed in air. Figure 5.20(b) shows the strength ratio of the soils placed on land and underwater and the sand content ratio, which was obtained in laboratory model tests. In the tests, the Hakata clay (w_L of 91%) and its mixture with sand were stabilized with ordinary Portland cement of 80 kg/m³, and then were placed underwater at 700 mm from the bottom. The figure shows that the unconfined compressive strength ratio of the underwater specimen is about 0.8 to 0.9 for the soil that has a quite small proportion of sand particles, but the strength ratio decreases with the sand particle content to about 0.4 to 0.8 when the sand particle content is about 90%.

When stabilized soil is placed underwater, soil separation can take place. Tang et al. (2002) carried out a field test to investigate soil separation in which stabilized Shimonoseki sand (sand particle content of 41% and silt and clay particles content of 35%) with slag cement of 120 kg/m³ was placed underwater. The upper part of the sediment was the clay, with a particle size less than 0.005 mm, while the lower part was sand with particle size exceeding 0.075 mm. Figure 5.20(c) shows the unconfined compressive strength of the stabilized soil together with that of the laboratory stabilized soil. The upper floating mud shows the strength increase with its unit weight irrespective of the segregation of the soil. The lower mud of segregated soil shows small strength, which was about 1,000 kN/m² smaller than for the non-segregated soil.

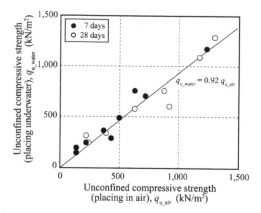

(a) Strength ratio of the stabilized soils placed on land and underwater.

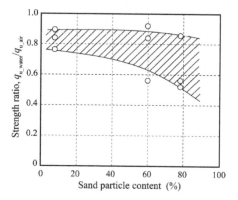

(b) Effect of sand particle content on the strength ratio of the stabilized soils placed on land and underwater.

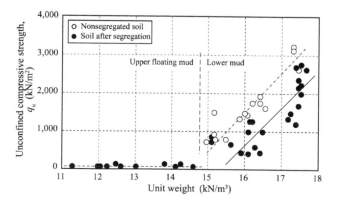

(c) Unconfined compressive strength of upper and lower part of stabilized soil sediment.

Figure 5.20 The effect of placement either on land or underwater, and the proportion of sand particle content on its unconfined compressive strength (Tang et al., 2002).

The test results have revealed that it is important not to cause soil separation during placing underwater, to assure the design strength.

4.3 Quality assurance

4.3.1 Shape of stabilized soil ground

The level of the stabilized soil ground is usually measured by a movable buggy equipped with a GPS system for placement on land and by a narrow multi-beam sonar for underwater placement, as already shown in Figure 5.17. The measured data are transmitted to the control room to draw the contour map of the site.

4.3.2 Strength of stabilized soil ground

Quality control should be performed during execution by appropriate methods in order to secure the prescribed quality requirement of the stabilized soil. The quality control of stabilized soil during execution is usually carried out by unconfined compression tests on samples taken at the energy dissipation cyclone (called as mold sample), and one after placement. A cone penetration test (JGS 1435) is sometimes carried out to investigate the short-term strength of stabilized soil ground. The correlation between the cone penetration resistance, q_c, and unconfined compressive strength, q_u, is established in advance by a series of laboratory mixing tests. The cone penetration resistance profile along the depth of the stabilized soil ground and the relationship between the cone penetration registance and the unconfined compressive strength are shown in Figures 3.62 and 3.63, where the Nagoya Port clay (w_L of 74.4% and w_P of 33.0%) was stabilized with ordinary Portland cement with various cement factors (Ministry of Transport, The Fifth District Port Construction Bureau, 1999).

4.3.3 Environmental impact during placement

A suitable placement technique should be selected to minimize the environmental impact on the surrounding area. Fine particles of stabilized soil are separated when the soil is not placed by the appropriate machine and procedure, which causes high turbidity in the water. Where the stabilized soil is placed underwater, the pH value of the water becomes high due to the chemical hydration of cement. Figure 5.21 shows the relationship between turbidity (Figure 5.21(a)) and the potential hydrogen, pH (Figure 5.21(b)), plotted against the distance from the deposition outlet (Tang et al., 2002). In the figure, not only the field case histories but also the laboratory model test results are plotted together. The case histories shown in the figure were performed by the pressured placement technique. Figure 5.21(a) clearly shows that high turbidity can be seen within about 1 m from the outlet, but the turbidity shows almost constant value ranging from 1 to 10 ppm for a distance of about 5 m from the outlet. A similar phenomenon can be seen in the pH value in Figure 5.21(b), where the pH value is very high close to the outlet – within about 1 m – but decreases rapidly with distance to an almost constant of about 8 at about 5 m from the outlet.

In natural conditions, the turbidity and pH value usually fluctuate, e.g. 1 to 10 ppm in turbidity and 8.0 to 8.5 in pH measured at the Tokoname Bay (Ministry of Transport, The Fifth District Port Construction Bureau, 1999).

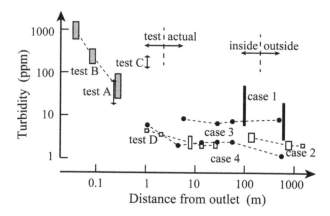

(a) Turbidity.

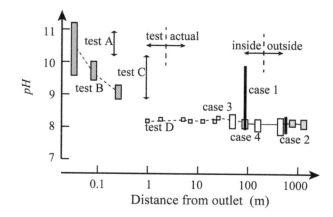

(b) Potential hydrogen (pH).

Figure 5.21 Turbidity and pH of water with distance from deposition outlet for stabilized soil placed underwater (Tang et al., 2002).

Figures 5.22 and 5.23 show the relationships between the suspended solids and the distance from revetment, which were measured in the field test at Nagoya Port (Ministry of Transport, The Fifth District Port Construction Bureau, 1999). The measured suspended solids, SS, value in the sea water outside the sea revetment (Figure 5.22) shows a maximum value of about 15 mg/L, which is almost the same as the background value there. This reveals that the placement of stabilized soil causes negligible influence to the marine environment outside revetment. The measured SS value in the sea water inside sea revetment (Figure 5.23), on the other hand, shows a maximum value of about 240 mg/L, which reveals that a countermeasure is necessary for the suspended solids.

Figures 5.24 and 5.25 show the relationship between the measured potential hydrogen, pH, during construction and that of the background in the field test at

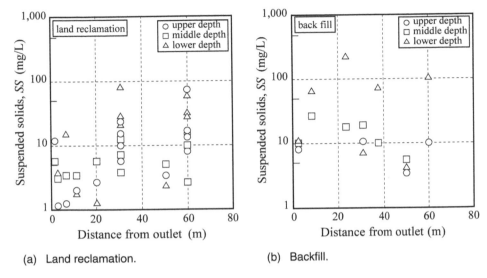

Figure 5.22 Measured suspended solids values in water outside a sea revetment (Ministry of Transport, The Fifth District Port Construction Bureau, 1999).

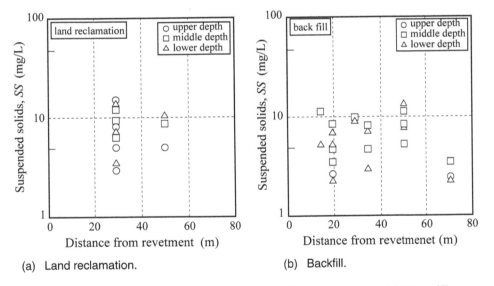

Figure 5.23 Measured suspended solid values in water inside a sea revetment (Ministry of Transport, The Fifth District Port Construction Bureau, 1999).

Nagoya Port (Ministry of Transport, The Fifth District Port Construction Bureau, 1999). The pH value during construction in water outside the revetment (Figure 5.24) shows a maximum difference of about 0.3 compared to the background, which is within the influence of the tide. This reveals that the placement of stabilized soil causes

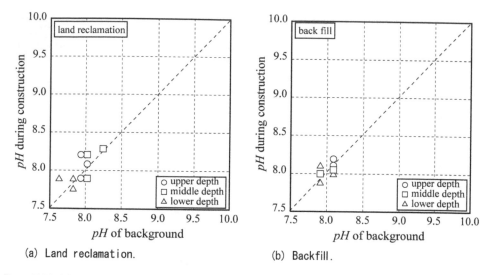

Figure 5.24 Measured pH values in water outside a sea revetment (Ministry of Transport, The Fifth District Port Construction Bureau, 1999).

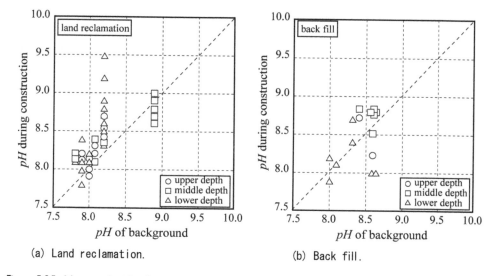

Figure 5.25 Measured pH value in water inside a sea revetment (Ministry of Transport, The Fifth District Port Construction Bureau, 1999).

negligible influence in the pH value to the water outside the sea revetment. The measured pH value for water inside the sea revetment (Figure 5.25), on the other hand, shows a maximum value of about 9.5, which reveals that a countermeasure is necessary to reduce the pH.

4.3.4 Water quality control

The water at a placement site contains a lot of suspended solids and is highly alkaline when cement is used as a binder. In Japan, the potential hydrogen, pH, value is specified as an environmental standard for defining marine pollution. The environmental standard for turbidity is not specified, but the Water Pollution Control Act specifies that the effluent standard for turbidity, is such that suspended solids should be lower than 200 ppm, corresponding to a turbidity lower than 125 ppm (Environment Agency, 2005).

The quality of the water should be monitored and controlled to the regulated levels before being allowed to flow out to a marine area. Figure 5.26(a) shows an example of a sedimentation basin at a spillway for the control of effluent, whose effective width, length, depth and sectional area were 33 m, 90 m, 7.6 m and 2,970 m^2, respectively, for sufficient precipitation of suspended solids (Satoh, 2003, 2004). In order to control the potential hydrogen, pH, of water to the regulated level, sulphuric acid, hydrochloric acid and carbon dioxide gas can be used. However, carbon dioxide gas has most often been used due to its low risk and easy handling. The water with high levels of suspended solids and which is highly alkaline is pumped to mixing tanks, where it is mixed with a polymer flocculating agent and carbon dioxide gas, and then it flows to the sedimentation basin. In the basin, precipitation of suspended solids, SS, in the water is accelerated by the polymer flocculating agent to the regulated level, i.e. SS lower than 60 mg/L. The pH is also controlled by the carbon dioxide gas to the regulated level of pH 8.4.

Figure 5.26(b) shows the amount of suspended solids and time history measured at the placement site and in the effluent in the Central Japan International Airport construction project (Satoh, 2003, 2004). In the figure, the highest values of the day are plotted. The SS value at the placement site is negligibly influenced by the progress of reclamation and has a mean value of about 37.8 mg/L which is lower than the regulation. The SS value at the effluent has a mean value of about 15.2 mg/L and is lower than the regulation.

Figure 5.26(c) shows the potential hydrogen, pH, and time history measured at the placement site and in the effluent in the Central Japan International Airport construction project (Satoh, 2003, 2004). The estimated value increases with the progress of reclamation to about a pH of 10 at 90% completion. The measured pH value at the placement site also increases with the progress of reclamation, to a peak of pH 9.8 at about 70% progress, but decreases with further reclamation. The pH value measured in the effluent is almost always lower than the regulation of 8.4.

4.3.5 Elution of hexavalent chromium (chromium VI) from stabilized soil

The elution of hexavalent chromium (chromium VI) from cement stabilized soils is specified by the environmental quality standards (Environment Agency, 2005) and the amount of hexavalent chromium should be measured before, during, and after execution and confirmed to be lower than the regulated values.

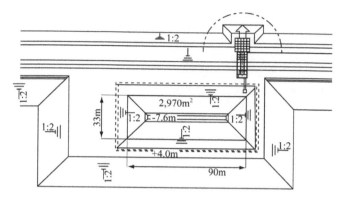

(a) Sedimentation basin at spillway.

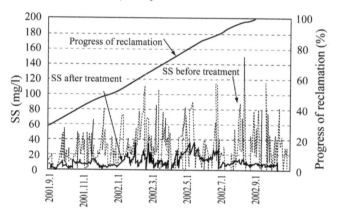

(b) Measured SS value inside revetment.

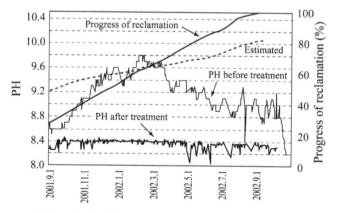

(c) Measured pH value inside revetment.

Figure 5.26 Water quality control and SS and pH time history in the Central Japan International Airport construction (Satoh, 2003, 2004).

REFERENCES

Environment Agency (2005) *Criteria for a Specific Operation of the Ground Storage Tank Outdoors using Deep Mixing Method (notification)*. (in Japanese).

Hashimoto (1999) Plug magic method. Inner pipe mixing solidification process. *Marine Voice 21*. Vol. 205, pp. 14–16 (in Japanese).

Hayano, K. & Kitazume, M. (2005) Strength variance within cement treated soils induced by newly developed pneumatic flow mixing method. *Proc. of the ACSE, Geo-Frontiers 2005*. CD-ROM.

Ikegami, N. (1999) On development of the K-DPM method. Mixing solidification processing transfer method in tube. *Marine Voice 21*. Vol. 207, pp. 24–26 (in Japanese).

Ishikawa, K. (2000) Pneumatic flow mixing method: Uniform Mixing Method. *Marine Voice 21*. Vol. 214, pp. 14–16 (in Japanese).

Iwata, H., Okumura, Y., Kawauchi, M., Satou, Y. & Takamatsu, H. (2000) Beneficial use of soft dredged soil by pneumatic flow mixing method, Development of drum mixing method. *Proc. of the 55th Annual Conference of the Japan Society of Civil Engineers*. V-158. (in Japanese).

Iwata, H., Okumura, Y., Kawauchi, M., Satou, Y., Saitoh, T. & Ishida, H. (2006) Development of manufacturing technique for high quality stabilized dredging soil using a compressed-air mixture pipeline with a drum-style mixer. *Proc. of Civil Engineering in the Ocean, Japan Society of Civil Engineers*. Vol. 22, pp. 951–956 (in Japanese).

Kitazume, M. & Hayano, K. (2005a) Strength scatter and its causes of Pneumatic Flow Mixing treated ground. *Report of the Port and Airport Research Institute*. Vol. 44, No. 2, pp. 57–81 (in Japanese).

Kitazume, M. & Hayano, K. (2005b) Strength property and variance of cement treated ground by pneumatic flow mixing method. *Proc. of the 6th International Conference on Ground Improvement Techniques*. pp. 377–384.

Kitazume, M. & Hayano, K. (2007) Strength property and variance of cement treated ground by pneumatic flow mixing method. *Ground Improvement Journal of ISSMGE*. Vol. 11, No.1, pp. 21–26.

Kitazume, M., Yoshino, N., Shinsha, H., Horii, R. & Fujio, Y. (2000) Field test on pneumatic flow mixing method for sea reclamation. *Proc. of the International Symposium on Coastal Geotechnical Engineering in Practice*. Vol. 1, pp. 647–652.

Kobayashi, K., Yoshida, G. & Sato, H. (2001) Quality assurance of pneumatic flow mixing method: Snake Mixer Method. *Proc. of the Annual Research Conference, Civil Engineering Research Institute for Cold Region, Ministry of Ministry of Land, Infrastructure, Transport and Tourism*. pp. 393–400 (in Japanese).

Ministry of Transport, The Fifth District Port Construction Bureau (1999) *Pneumatic Flow Mixing Method*. Yasuki Publishers. 157p. (in Japanese).

Mori, Y. (2000) Pneumatic flow mixing method: Balance mixing method. *HEDORO*. Vol. 79, pp. 43–49 (in Japanese).

Mori, Y. (2001) Recycle of dredged soil/ Pneumatic flow mixing method, balance mixing method. *Marine Voice 21*. Vol. 218, pp. 21–23 (in Japanese).

Ogawa, H. (1999) Development of snake mixing method. *Annual Research Meeting, Shikoku Branch, Japan Society of Civil Engineers*. VI-7, pp. 386–387 (in Japanese).

Oota, M. & Sakamoto, A. (2008) Development and application of plug magic method: Beneficial use of dredged soil for construction material. *Journal of the Japanese Society of Soil Mechanics and Foundation Engineering, Tsuchi to Kiso*. Vol. 56, No. 12, pp. 46–47 (in Japanese).

Sakamoto, A. (1998) In-pipe mixing solidification method PLUG-MAGIC method. *HEDORO*. Vol. 73, pp. 41–46 (in Japanese).

Sasaki, Y. (1999) Mud solidification air-transfer process. LMP process. *Marine Voice 21*. Vol. 205, pp. 17–19 (in Japanese).

Sato, Y. & Hayashida, H. (2000) Field report on W-pipe mixing method at Fushikitoyama Port. *Annual Technical Report, Wakachiku Construction Co., Ltd.* Vol. 9, pp. 27–37 (in Japanese).
Satoh, A. (2001) Balance Mixing Method, Pneumatic flow mixing method. *Kensetsu Kikai, Japan Industry Publishers.* Vol. 37, pp. 46–52 (in Japanese).
Satoh, T. (2003) Application of pneumatic flow mixing method to Central Japan International Airport Construction. *Journal of the Japan Society of Civil Engineers.* No. 749/6-61, pp. 33–47 (in Japanese).
Satoh, T. (2004) Development and application of pneumatic flow mixing method to reclamation for offshore airport. *Technical Note of the Port and Harbour Research Institute.* No. 1076, 81p. (in Japanese).
Shima, M. & Hashimoto, F. (1998) Plug magic method for effective weak dredged soil application. In-tube mixing solidification treatment method. *Kouwan, The Japan Port & Harbour Association.* Vol. 75, No. 2, p. 51 (in Japanese).
Sumoto, T. (2000) Field Test on Pneumatic flow mixing method (Balance Mixing Method). *Proc. of the 55th Annual Conference of the Japan Society of Civil Engineers.* pp. 582–583 (in Japanese).
Tang, Y. X., Miyazaki, Y., Ochiai, H., Yasufuku, N. & Omine, K. (2002) Environmental impacts on seawater due to casting cement treated soil underwater. *Journal of Geotechnical Engineering, Japan Society of Civil Engineers.* No.708/3-59, pp. 211–220 (in Japanese).

Chapter 6

Geotechnical design of stabilized soil ground

I INTRODUCTION

The pneumatic flow mixing method produces stabilized soil for reclaimed ground whose strength and compressibility values are higher than those of ordinary soil. The application of the method includes the decrease in active earth pressure, increase in passive earth pressure, shallow reinforcement, and improving dynamic response, *etc.* as already shown in Figure 1.16. The geotechnical design of stabilized soil ground is carried out in a similar manner to ordinary soil, such as clay and sand. As the physical and mechanical properties of stabilized soil are affected by many factors, as explained in Chapter 2, and have relatively large variations in many cases, the design strength for each application should be carefully determined by taking into account these effects.

The technical standard for the geotechnical design of improved ground by the pneumatic flow mixing method was first established in 1999 by Ministry of Transport (Ministry of Transport, 1999), which was revised in 2007 (Ministry of Land, Infrastructure, Transport and Tourism, 2007). The Ports and Harbours Association of Japan published the standard and commentaries of the original Japanese version (The Ports and Harbours Association of Japan, 1999, 2007) and the Overseas Coastal Area Development Institute of Japan published the English version (The Overseas Coastal Area Development Institute of Japan, 2002, 2009).

In this chapter, the determination of the design strength of stabilized soil is briefly introduced at first, then the geotechnical design procedure for the earth pressure and the bearing capacity of stabilized soil ground are introduced. Finally, the soil volume design is explained, which is a typical issue of the pneumatic flow mixing method.

2 DESIGN STRENGTH

2.1 Relationships of laboratory strength, field strength and design strength

Whatever the type of application and the function of stabilized soil ground, it is important to discuss the strengths of laboratory and field stabilized soils. As described in Chapter 2, the physical and mechanical properties of stabilized soil are affected by many factors such as the soil properties; the type and quantity of binder; mixing and placement conditions; and curing conditions. The effects of these factors are quite complex, making it difficult to directly estimate by laboratory mix test only the strength of soil stabilized in the field.

Field mixing conditions and curing conditions are quite different from standard laboratory mix test conditions, and the strength of field stabilized soil is not always the same as that estimated by the laboratory mix test. Field stabilized soil has a relatively large strength variability, even if the execution is carried out with an established mixing system and procedure and with the best care. The frequency distributions of laboratory stabilized soil strength and field stabilized soil strength can be schematically shown in Figure 6.1, in which the $\overline{q_{ul}}$ and $\overline{q_{uf}}$ are the average unconfined compressive strength of laboratory stabilized soil and field stabilized soil, respectively. Usually field stabilized soil has a smaller average strength and larger strength deviation than those of laboratory stabilized soil. Therefore, the frequency distribution of field stabilized soil strength is gentle slope shape, and its plot has a smaller peak than the laboratory stabilized soil. In Figure 6.1 the design strength, q_{uck}, is also plotted. The magnitude of design strength is obtained in the geotechnical design and depends on the purpose and application of the stabilized soil and the type, function and importance of the superstructure on it. A large part of field stabilized soil strength should be larger than the design strength, while the percentage of field stabilized soil strength larger than the q_{uck} can be defined as 'probability'. The design strength, field strength and laboratory strength are formulated as Equation 6.1. The K is a coefficient related to the probability. When the frequency distribution of field soil strength is assumed as a normal distribution, the relationship between the K and the probability is obtained as in Table 6.1.

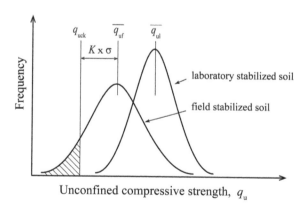

Figure 6.1   Schematic of the frequency distributions of field stabilized soil strength and laboratorystabilized soil strength.

Table 6.1   Relationship between K value and probability in the case of a normal distribution for stabilized soil strength.

Probability	30%	25%	20%	15%	10%	5%
K	0.524	0.674	0.841	1.036	1.281	1.644

$$q_{\text{uck}} \leq \overline{q_{\text{uf}}} - K \cdot \sigma \qquad\qquad (6.1)$$
$$\overline{q_{\text{uf}}} = \lambda \cdot \overline{q_{\text{ul}}}$$

where

K: coefficient

q_{uck}: design standard strength (kN/m^2)

$\overline{q_{\text{uf}}}$: average unconfined compressive strength of field stabilized soil (kN/m^2)

$\overline{q_{\text{ul}}}$: average unconfined compressive strength of laboratory stabilized soil (kN/m^2)

σ: standard deviation of the field strength (kN/m^2)

λ: ratio of $\overline{q_{\text{uf}}}/\overline{q_{\text{ul}}}$.

2.2 Design flow for field and laboratory stabilized soil strengths and mixing condition

Figure 6.2 shows the design flow for determining the mixing condition of stabilized soil. As the first part of the procedure, the design strength of stabilized soil, q_{uck},

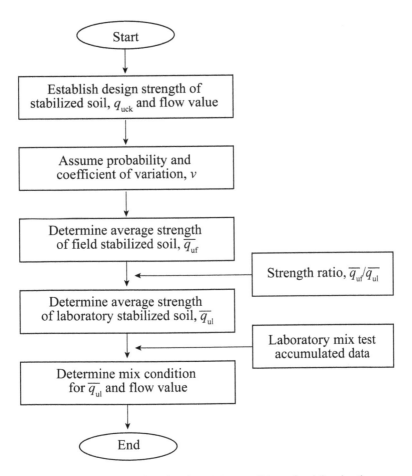

Figure 6.2 Design flow for the mixing condition of stabilized soil.

is determined by the geotechnical design, which will be explained later. Then, the magnitude of probability should be assumed, which is the degree to which the field stabilized soil strength exceeds the specified design strength. An appropriate value of probability depends on the purpose and application of stabilized soil, and the type, function and importance of the superstructure on it. In many construction projects, a probability of 75% has often been applied where the K value is determined as 0.67 by assuming a normal distribution curve for the field stabilized soil strength. According to the accumulated case histories, the coefficient of variation of field stabilized soil strength and the strength ratio, $\overline{q_{uf}}/\overline{q_{ul}}$, are usually assumed to be 0.35 and 0.7, respectively in Japan. After these assumptions and procedures, the average unconfined compressive strengths of field stabilized soil strength, $\overline{q_{uf}}$, and of the laboratory stabilized soil strength, $\overline{q_{ul}}$, are determined. The mixing condition for achieving the laboratory stabilized soil strength is usually determined by the laboratory mix test and/or accumulated data.

3 GEOTECHNICAL DESIGN

The target strength of the method is dependent upon the purpose of the application, but in many cases is adopted as 100 to 200 kN/m^2 in q_u. As described in Chapter 3, the stabilized soil shows brittle characteristics, where there is a sharp peak strength, a small axial strain at failure, and a small residual strength. The failure pattern of stabilized soil is different from that of ordinary soil ground, which is influenced by the shape and location of stabilized soil and the external loading conditions (Kitazume et al., 1996, 1997). Though these characteristics are different from ordinary clay, stabilized soil is assumed to be the same as ordinary cohesive soil having ductile characteristics in the current design standard. In the design methods for bearing capacity and earth pressure, the bearing capacity is almost the same as that applying to ordinary soils. In the following sections, the current design methods are briefly introduced.

3.1 Earth pressure of stabilized soil ground with infinite width

The formulations below are provided under the assumption that a stabilized soil ground is a horizontally stratified ground, having infinite width and uniform properties.

3.1.1 Earth pressure before hardening

The stabilized soil is in a liquid state at mixing and then attains a solid state with the progress of chemical hydration. The stability of the retaining structure should be designed by incorporating the earth pressure of not only the solid state condition but also the liquid state condition. According to the previous research (e.g. Horiuchi et al., 1992; Kawasaki et al., 1992; Kitazume & Yamamoto, 1997), the earth pressure of the liquid state condition can be assumed to be as the same as fluid pressure, which is calculated by Equation 6.2.

$$p_t = \gamma_t \cdot h \qquad (6.2)$$

where
h: height of liquid state stabilized soil layer (m)
p_t: earth pressure of liquid state stabilized soil (kN/m^2)
γ_t: unit weight of liquid state stabilized soil (kN/m^3).

3.1.2 Earth pressure after hardening

(a) Earth pressure at rest
The earth pressure at rest of stabilized soil, p_0', is calculated by Equation 6.3. The magnitude of the coefficient of earth pressure at rest, K_0, is provided as 0.15 to 0.2 (Figure 3.35) (Kitazume & Yamamoto, 1997).

$$p_0' = K_0 \left(\Sigma \gamma' \cdot h + w \right) \qquad (6.3)$$

where
h: thickness of stabilized soil layer (m)
K_0: coefficient of earth pressure at rest
p_0': effective earth pressure at rest coefficient
w: surcharge per unit area at ground surface (kN/m^2)
γ': effective unit weight of stabilized soil (kN/m^3).

(b) Static active earth pressure
The static active earth pressure of stabilized soil is calculated by Equation 6.4. The cohesion, c, in the equation is usually assumed as $q_u/2$. The cohesion of stabilized soil is so large in many cases that the calculated active earth pressure is a negative value in a large area, especially at a shallow depth. In the calculation for stability problems, the active earth pressure is usually evaluated as 0 when it is a negative value.

$$p_a' = \Sigma \gamma' \cdot h + w - 2c \qquad (6.4)$$

where
c: cohesion of stabilized soil (kN/m^2)
h: thickness of stabilized soil layer (m)
p_a': active earth pressure (kN/m^2)
w: surcharge per unit area at ground surface (kN/m^2)
γ': effective unit weight of stabilized soil (kN/m^3).

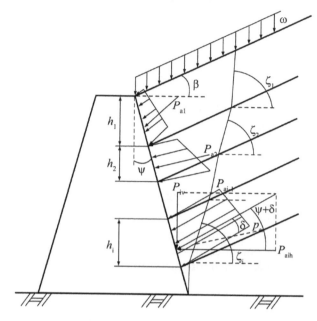

Figure 6.3 Dynamic active earth pressure calculation.

(c) Dynamic active earth pressure

The dynamic active earth pressure of stabilized soil is calculated by Equation 6.5 (see Figure 6.3) (Ministry of Land, Infrastructure, Transport and Tourism, 2007).

$$p_{ai} = \left\{ \frac{\left(\sum \gamma_i h_i\right) \cos(\psi - \beta)}{\cos \psi} + w \right\} \frac{\sin(\zeta_i - \phi_i + \theta) \cos(\psi - \zeta_i)}{\cos \theta \cos(\psi - \zeta_i + \phi_i + \delta) \sin(\zeta_i - \beta)}$$

$$- \frac{c_i \cos(\psi - \beta) \cos \phi_i}{\cos(\psi - \zeta_i + \phi_i + \delta) \sin(\zeta_i - \beta)}$$

$$\zeta_i = \frac{\psi + \phi_i - \mu_i + 90^0}{2}$$

$$\mu_i = \tan^{-1} \frac{B_i C_i + A_i \sqrt{B_i^2 - A_i^2 + C_i^2}}{B_i^2 - A_i^2}$$

$$\theta = \tan^{-1} k_h \text{ or } \tan^{-1} k_h'$$

$$A_i = \sin(\delta + \beta + \theta)$$

$$B_i = \sin(\psi + \phi_i + \delta - \beta) \cos \theta - \sin(\psi - \phi_i + \theta) \cos(\delta + \beta)$$

$$+ \frac{2c_i \cos(\psi - \beta) \cos \phi_i \cos(\delta + \beta) \cos \theta}{\dfrac{\left(\sum \gamma_i h_i\right) \cos(\psi - \beta)}{2 \cos \psi} + w}$$

$$C_i = \sin(\psi + \phi_i + \delta - \beta) \sin \theta + \sin(\psi - \phi_i + \theta) \sin(\delta + \beta)$$

$$- \frac{2c_i \cos(\psi - \beta) \cos \phi_i \sin(\delta + \beta) \cos \theta}{\dfrac{\left(\sum \gamma_i h_i\right) \cos(\psi - \beta)}{2 \cos \psi} + w}$$

(6.5)

where
c_i: cohesion of soil of the i-th layer (kN/m^2)
h_i: thickness of the i-th layer (m)
k_h: horizontal seismic coefficient
k'_h: apparent horizontal seismic coefficient
p_{ai}: dynamic active earth pressure of the i-th layer (kN/m^2)
w: surcharge per unit area at ground surface (kN/m^2)
β: angle of back fill to the horizontal (°)
δ: angle of wall friction (°)
ϕ_i: internal friction angle of the i-th layer
γ_i: unit weight of the i-th layer (kN/m^3)
θ: resultant seismic coefficient angle (°)
ψ: angle of wall to the vertical (°)
ζ_i: angle of failure surface on the i-th layer to the horizontal (°).

3.2 Earth pressure of stabilized soil ground with a finite width

In many cases, the width of stabilized soil ground is not wide enough to adopt Equations 6.4 and 6.5 for calculating the static and dynamic active earth pressures. And in the condition where the stabilized soil portion is not of uniform shape and stratification, Equations 6.4 and 6.5 are not adopted. For these cases, a slice calculation method should be used to calculate the active earth pressure of stabilized soil and surrounding soils. In the slice method, the active earth pressure is calculated by load equilibrium of the weight, buoyancy load, shear strength mobilized, and seismic forces along an assumed linear slip surface as illustrated in Figure 6.4.

In the case where the thickness of stabilized soil ground is limited, the earth pressure is calculated by the slice method as Equations 6.6.

$$p \cdot \cos \delta = \frac{\Sigma W_i \cdot k_h + \dfrac{-c \cdot l_i \cdot \sec \alpha + W'_i \cdot (\tan \alpha - \tan \phi)}{1 + \tan \alpha \cdot \tan \phi}}{1 + \dfrac{\tan \alpha - \tan \phi}{1 + \tan \alpha \cdot \tan \phi} \cdot \tan \delta} \tag{6.6}$$

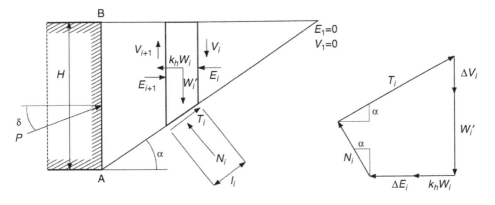

Figure 6.4 Earth pressure calculation by slice method.

where
c: cohesion of stabilized soil (kN/m^2)
k_h: horizontal seismic coefficient
l_i: length of sliced portion of slip surface
p: total of earth pressure
W_i: weight of sliced portion
W_i': effective weight of sliced portion (weight after subtracting buoyancy)
α: angle of slip surface
δ: friction angle at wall surface
ϕ: internal friction angle.

In the calculations, the three failure modes are assumed as shown in Figure 6.5 (Tsuchida & Egashira, 2004): (I) is a linear failure surface passing through the stabilized soil, (II) is a combined failure surface consisted of a linear failure surface passing through surrounding ground and a tensile crack developed in the stabilized soil, and (III) is a linear failure surface passing along the boundary surface of stabilized soil. Mode (I) can be critical for a case where the stabilized soil has relatively small strength and/or a small width. In mode (II), a combined failure mode with shear and tensile failures is assumed, which is observed in several previous studies (Kitazume et al., 2003a, 2003b). Mode (III) can be critical for a case where the stabilized soil has a relatively

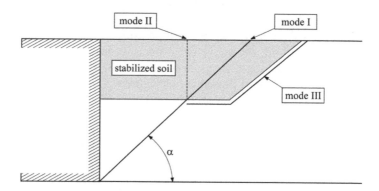

(a) Assumed failure modes for the earth pressure calculation of stabilized soil.

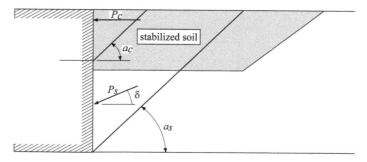

(b) Wall friction for the earth pressure calculation of stabilized soil.

Figure 6.5 The slice method used for active earth pressure calculation (Tsuchida & Egashira, 2004).

large strength and/or a large thickness. The active earth pressure is determined as the maximum value among the calculations by three failure modes. Though the magnitude of wall friction is still unclear, in the current design, the friction component is taken into account and the wall friction angle, δ, is assumed to be 0.26 radian (15°), while the cohesive component is neglected.

3.3   Bearing capacity of stabilized soil ground

It is widely recognized that stabilized soil ground shows various failure modes, depending upon the strength, the extent and the shape of the stabilized soil ground, and also upon the external load condition, which are shear failure, bending failure and tensile failure (Kitazume et al., 2003a, 2003b, Kitazume & Maruyama, 2007). The bearing capacity of the stabilized soil ground is influenced by these failure modes. However, no design procedure on the bearing capacity of stabilized soil ground has been proposed yet which incorporates the effect of failure mode. In the case of the stabilized soil ground having a sufficiently large spatial and sectional extent, the bearing capacity of stabilized soil ground is evaluated by conventional bearing capacity theory for clay ground, as shown in Equation 6.7 (Ministry of Transport, 1999). The appropriate magnitude of the bearing capacity factors should be determined by considering accumulated research results. In the case where the stabilized soil ground having a sufficiently large spatial extent but limited thickness – a sort of slab –, the bearing capacity is calculated by the Winkler theory.

$$q_f = \frac{1}{Fs} \left(\frac{1}{2} \gamma \cdot B \cdot N_\gamma + c_{ub} \cdot N_c + q \cdot (N_q - 1) \right) + q \qquad (6.7)$$

where
B: width of superstructure (m)
c_{ub}: undrained shear strength of soil beneath improved ground (kN/m^2)
Fs: safety factor
N_c: bearing capacity factor of soil beneath improved ground
N_q: bearing capacity factor of soil beneath improved ground
N_γ: bearing capacity factor of soil beneath improved ground
q: effective overburden pressure at bottom of improved ground (kN/m^2)
q_f: bearing capacity of soil beneath improved ground (kN/m^2)
γ: unit weight of soil beneath improved ground (kN/m^3).

3.4   Liquefaction of stabilized soil

In the case where stabilized soil is applied to the purpose of liquefaction prevention, according to the previous researches (Zen et al., 1987), a soil having a q_u exceeding about 100 kN/m^2 can be assumed not to liquefy. It can be concluded that the stabilized soil does not liquefy as long as its q_u value exceeds about 100 kN/m^2.

3.5   Soil volume design

In pneumatic flow mixing projects, original soil is excavated and an amount of water and cement are added and mixed to obtain stabilized soil for the required design

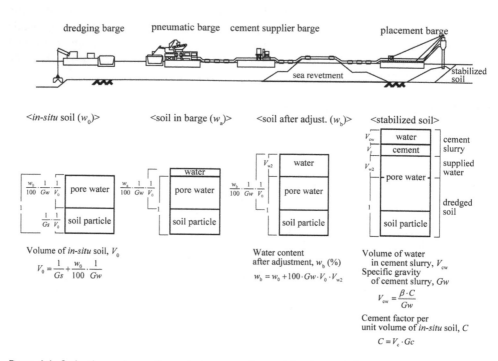

Figure 6.6 Soil volume change through the excavation, stabilization and placement processes (Ministry of Transport, The Fifth District Port Construction Bureaus, 1999).

characteristics. Figure 6.6 illustrates the soil volume change through the excavation, stabilization and placement processes (Ministry of Transport, The Fifth District Port Construction Bureau, 1999). In order to obtain the required volume of dredged soil for the volume of reclaimed ground, the change of soil volume should be evaluated precisely. The volume ratio between the soil to be dredged and the soil on the transporting barge, N_{vol}, can be calculated by Equation 6.8a, and that between the soil to be dredged and the stabilized soil, N'_{vol}, can be calculated by Equation 6.8b. Therefore, the volume of stabilized soil after stabilization is $N_{\text{vol}} \times N'_{\text{vol}}$ times larger than that at the excavation site.

$$N_{\text{vol}} = 1 + V_{\text{w2}} + V_{\text{c}} + V_{\text{cw}} = \frac{\dfrac{w_{\text{b}} \cdot G_{\text{s}}}{100} + G_{\text{w}}}{\dfrac{w_0 \cdot G_{\text{s}}}{100} + G_{\text{w}}} + \frac{C}{G_{\text{c}}} + \frac{W/C \cdot C}{G_{\text{w}}} \tag{6.8a}$$

$$N'_{\text{vol}} = \frac{\dfrac{w_{\text{b}} \cdot G_{\text{s}}}{100} + G_{\text{w}} + C\left(\dfrac{w_0 \cdot G_{\text{s}}}{100} + G_{\text{w}}\right)\left(\dfrac{1}{G_{\text{c}}} + \dfrac{W/C}{G_{\text{w}}}\right)}{\dfrac{w_{\text{a}} \cdot G_{\text{s}}}{100} + G_{\text{w}}} \tag{6.8b}$$

where
C: cement factor (kg/m^3)
G_c: specific gravity of cement
G_s: specific gravity of soil particle
G_w: specific gravity of water
w_0: water content of soil at exavation site (%)
w_a: water content of soil on transporting barge (%)
w_b: water content of soil after adjustment (%)
V_c: volume of cement (m^3)
V_{cw}: volume of water in cement slurry (m^3)
V_{w2}: volume of water added on transporting barge (m^3)
W/C: water to cement ratio of cement slurry.

REFERENCES

Horiuchi, S., Taketsuka, M., Odawara, T. & Kawasaki, H. (1992) Fly-ash slurry island: I Theoretical & experimental investigations, *Journal of Materials in Civil Engineering. ASCE*, pp. 117–133.

Kawasaki, H., Horiuchi, S., Akatsuka, M. & Sano, S. (1992) Fly-ash slurry island: II Construction in Hakucho Ohashi Project, *Journal of Materials in Civil Engineering. ASCE*, pp. 134–152.

Kitazume, M. & Maruyama, K. (2007) Internal Stability of Group Column Type Deep Mixing Improved Ground under Embankment Loading. *Soils and Foundations*. Vol. 47, No. 3, pp. 437–455 (in Japanese).

Kitazume, M. & Yamamoto, H. (1997) Failure of Cement Treated Fly Ash Ground. *Report of the Port and Harbour Research Institute*. Vol. 36, No. 1, pp. 1–23 (in Japanese).

Kitazume, M., Hayano, K. & Hashizume, H. (2003a) New limit equilibrium method for seismic stability of cement treated clay ground. *Proc. of the 12th Asian Regional Conference on Soils Mechanics and Geotechnical Engineering*. Vol. 1, pp. 299–302.

Kitazume, M., Hayano, K. & Hashizume, H. (2003b) Seismic stability of cement treated ground by titling and dynamic shaking table tests. *Soils and Foundations*. Vol. 43, No. 6, pp. 125–140.

Kitazume, M., Miyajima, S., Ikeda, T. & Doncho Karastanev (1996) Bearing Capacity of Improved Ground with Column Type DMM. *Proc. of the 2nd International Conference on Ground Improvement Geosystems*. Vol. 1, pp. 503–508.

Kitazume, M., Yamamoto, H. & Oishi, T. (1997) Failure Pattern and Earth Pressure of Cement Treated Fly Ash. *Proc. of the 12th International Symposium on Management and Use of Coal Combustion By-Products*. pp. 58–63.

Ministry of Land, Infrastructure, Transport & Tourism (2007) *Technical Standards for Port and Harbour Facilities*. The Ports and Harbours Association of Japan. Vol. 2, pp. 672–762 (in Japanese).

Ministry of Transport (1999) *Technical Standards for Port and Harbour Facilities*. The Ports and Harbours Association of Japan. pp. 525–536 (in Japanese).

Ministry of Transport, The Fifth District Port Construction Bureau (1999) *Pneumatic Flow Mixing Method*. Yasuki Publishers. 157p. (in Japanese).

The Overseas Coastal Area Development Institute of Japan (2002) *Technical Standards and Commentaries for Port and Harbour Facilities in Japan (English Version)*.

The Overseas Coastal Area Development Institute of Japan (2009) *Technical Standards and Commentaries for Port and Harbour Facilities in Japan (English Version)*.

The Ports & Harbours Association of Japan (1999) *Technical Standards and Commentaries for Port and Harbour Facilities in Japan (Japanese Version)*. (in Japanese).

The Ports & Harbours Association of Japan (2007) *Technical Standards and Commentaries for Port and Harbour Facilities in Japan (Japanese Version)*. (in Japanese).

Tsuchida, T. & Egashira, K. (2004) *The Lightweight Treated Soil method – New Geomaterials for Soft Ground Engineering in Coastal Areas -*. A.A.Balkema Publishers, 120p.

Zen, K., Yamazaki, H., Watanabe, A., Yoshizawa, H. & Tamai, A. (1987) Study on a reclamation method with cement-mixed sandy soils – Fundamental characteristics of treated soils and model tests on the mixing and reclamation. *Technical Note of the Port and Harbour Research Institute*. No. 579, 41p. (in Japanese).

Japanese laboratory mix test procedure

1 INTRODUCTION

The shear strength of stabilized soil is considered the most important geotechnical characteristic; it leads to the improved stiffness, homogeneity and long term stability of stabilized soil. In general, the shear strength of stabilized soil is influenced by many factors, including the characteristics of soil (water content, organic matter content, etc.); non-uniformity of soil (due to complex natural soil structure); type and amount of binder; curing period and temperature; and the degree of mixedness (Babasaki et al., 1996). Hence, it is difficult to predict precisely the strength of soil to be stabilized in the field solely by soil investigations prior to mixing. In order to determine the mix design for actual production it is very important to perform a laboratory mix test which examines the unconfined compressive strength of stabilized soils prepared in the laboratory, q_{ul}, by changing the type and amount of binder, curing time, and water cement ratio. This mix design process also contributes to quality control at the construction site. It is important to recognise that the strength of laboratory mixed stabilized soil, q_{ul}, is not always same as the strength of field mixed stabilized soil, q_{uf}. This knowledge may prevent troubles arising at the construction site. The strength of laboratory mixed stabilized soil is influenced by the procedure of making and curing stabilized soil. According to a recent questionnaire survey regarding protocols for laboratory mix test procedures, molding methods, and curing conditions exhibit notable international differences (Kitazume et al., 2009).

In this Appendix, a procedure for making and curing a stabilized soil specimen is introduced which is frequently applied in Japan to obtain the mixing condition required to assure the target strength, and to develop new binder. This procedure conforms to the Japanese Geotechnical Society Standard (Japanese Geotechnical Society, 2009).

2 TESTING EQUIPMENT

2.1 Equipment for making specimen

2.1.1 Mold

The standard mold size is 50 mm in diameter and 100 mm in height. However, depending on the soil characteristic, the specimen diameter may be varied. In the case of clayey or sandy soil without gravels, and when the amount of soil is limited, a diameter less

(a) Disposable plastic mold. (b) Disposable metal mold.

Figure A.1 Standard-sized lightweight mold used for testing stabilized soil in the laboratory.

than 50 mm has been used. Conversely, if the soil contains a large amount of gravels or decayed plants, a diameter larger than 50 mm can be accepted. In both cases, the height of the specimen is set to be 2.0 to 2.5 times the diameter.

The material for the mold is usually either cast iron, plastic, or tin. The latter two types of mold are referred to as lightweight molds and are popular choices today. The merits of lightweight molds are that they are easy to tap against the surface of a table or floor to remove air bubbles and it is easy to remove the specimen from the mold. Also, the specimen can be cured in the mold without the risk of the mold rusting. Figure A.1 shows photos of a standard-sized lightweight molds, 100 mm in height and 50 mm in diameter. Splittable cast-iron molds are also available in various sizes based on JIS A 1132 (Japanese Industrial Standard, 2006).

2.1.2 Mixer

A mixer should be capable of mixing soil and binder uniformly. An electric mixer consisting of three basic parts: motor, stirring blades, and mixing bowl is specified in the Japanese Geotechnical Society standard, because the electric mixer is suitable for most types of soil: clayey, organic, and sandy soils in most cases. Figure A.2 shows an example of electric mixer which is often used in Japan. The capacity of bowls ranges from 5,000 to 30,000 cm^3. Different types of mixing paddles are available, as shown in Figure A.3, but for most cases a hook type is preferred for uniform mixing. In this particular soil mixer, the paddle revolves at 120 to 300 rpm with planetary motions of 30 to125 rpm. The stand of the mixer enables the raising and lowering of the bowl during mixing.

2.1.3 Binder mixing tool

When binder in slurry form is to be used, use mixing bowl (typically a metal bowl) and a rubber spatula or spoon to mix the binder and water.

Figure A.2 Electric mixer.

(a) Hook type.   (b) Flat type.   (c) Whipper type.

Figure A.3 Examples of mixing paddles used in electric mixer.

2.2 Soil and binder

2.2.1 *Soil*

For a laboratory mix test for an actual construction purpose, it is a basic principle to collect soil samples from all soil layers to be stabilized. In order to collect soil samples from deeper layers, a thin-walled sampler is typically used. Sampled soil should be stored at its natural water content. The soil samples are classified based on observation records taken at source, and soil testing results. Natural water content, consistency limits, organic matter content, pH, and grain size distribution are good indices for their classification (see Chapter 2). The soil samples are separated into the identified layers. However, a soil sample from a thick layer is sometimes further divided into sublayers to take variation in water content into consideration. Conversely, in a case where a layer is thin and its soil characteristics are similar to those of its neighboring layer, these layers are combined, to reduce testing complexity. Each grouped soil sample is sieved through a 9.5 mm sieve. In a case where the diameter of the mold used is less than 50 mm, the soil sample is sieved through an appropriate size sieve so that the maximum grain size of the sieved sample should be less than 1/5th of the inner diameter of the mold. While sieving, large obstacles such as shells and plants should be removed. If it is clearly found that the grain size is less than 1/5th of the inner diameter of mold and the sample does not contain any obstacles, this procedure can be skipped. Then, each grouped soil sample is stirred by a mixer and its water content is measured. If it is considered that the water content of the soil sample has been changed during the process of sampling, transportation, and storage, the water content of the soil sample should be adjusted to its natural water content.

The required amount of soil sample is about 500 g for a standard-sized specimen. The total number of specimens to be tested is determined by the variations in binder types, binder factor (or binder content), curing period (curing time), and other construction control values (such as the influence of water/binder ratio), or a combination thereof. Three or more specimens should be prepared for each mixing condition and curing period. It is desirable to have an extra amount of each soil sample, in case there is a need to conduct follow-up tests or repeat tests (due to procedural errors).

Note: The sampling strategy mentioned above is applicable for mechanical mixing with vertical rotary shafts and blades. For shallow mixing techniques, or a chainsaw-type deep mixing system which involve the vertical movement of the soil–binder mixture in the actual production, soil samples may be prepared to simulate the *in situ* mixing condition by combining the soils taken from different layers according to the weighted average.

2.2.2 Binder

The quality of binder should be stringently assured. In general, it is desirable to use fresh binder for the test. However, if the use of aged binder is unavoidable, it should be inspected thoroughly for any quality degradation. For instance, degraded cement becomes grainy. The binder form in the mixing test is roughly divided between the slurry form and the powder or granular form. Chemical additives are sometimes used together with binder, which provide specific effects, such as accelerating or decelerating the rate of hardening. For instance, retarding chemical additives may be used for easing the process of overlapping stabilized soil columns in the case of the deep mixing method.

The required amount of binder is determined by the binder factor (or binder content) and the number of specimens. Similar to the required amount of soil sample, it is desirable to have an extra amount of binder. Tap water is generally used to make binder slurry. However, seawater may be used for marine construction.

3 MAKING AND CURING OF SPECIMENS

3.1 Mixing materials

The optimal duration to mix soil and binder varies due to many factors, such as the type and amount of soil, the type and amount of binder, and the consistency of the soil–binder mixture. The JGS standard specifies that the binder should be mixed with the soil thoroughly to achieve a uniform mixture and notes that the about 10 minutes is the ordinary practice, and 10 minutes is accepted as standard. When the mixing duration is too long, it becomes difficult to remove air bubbles from stabilized soil in a mold, since the stabilized soil may begin to harden.

Note: It is desirable to suspend the mixing after about 5 minutes, to detach the mixing bowl from the mixer, and to pour the stabilized soil in the mixing bowl and that adhered to the stirring blades to another container using a rubber spatula, to mix it briefly by hand, then to return it to the mixing bowl, and to restart to mix it by the mixer for another 5 minutes. Another option is to suspend the mixing every two minutes and to mix the soil in the mixing bowl by hand. These procedures provide uniform mixing of the soil, including the soil stuck to the mixing bowl and blades.

Where a slurry form binder is used, splashing of the slurry may occur when starting the mixer right after pouring the binder slurry onto the soil in the mixing bowl. It is desirable to mix the soil and the slurry by hand briefly before starting the mixer.

3.2 Making a specimen

A thin layer of grease may be applied to the inner surface of the mold to allow easy removal of the specimen after curing. Then the mold is filled with stabilized soil in three separate layers. After placing each stabilized soil layer in the mold, air bubbles should be removed. Typical methods for removing air bubbles are (1) lightly tapping the mold against a table or a concrete floor (Figure A.4), (2) hitting the mold with mallet, and (3) subjecting the mold to vibration. The air removing procedure is terminated once air bubbles are no longer found on the soil surface.

In general, it is hard to remove air bubbles from stabilized soil with a low consistency. Also, some stabilized soils decrease in volume over time, resulting in insufficient specimen height. To assure the proper specimen height, a sheet of hard polymer film, 10 to 15 mm taller than the mold height, is placed around the inner perimeter of mold so that stabilized soil can be filled above the top edge of the mold and be sealed by

Figure A.4 Tapping technique in molding procedure, used to release trapped air bubbles.

Figure A.5 Sealing the soil sample in the mold with plastic film.

sealant as shown in Figure A.5. The hard polymer film also functions to protect the specimen when it is removed from the mold.

The water content of stabilized soil is measured for each mixing bowl. By comparing the water content before and after mixing, any mistakes in material amounts can be spotted in the early experimental stage.

Some stabilized soils become hard quickly, making the removal of air bubbles difficult. In such a case, the stabilized soil mixture should be put in molds as quickly as possible by increasing the number of personnel and/or dividing the task into several batches, and reducing the quantity of material in each batch.

Sandy soil and binder sometimes separate easily during mixing and placing into molds, especially when slurry form binder is used. This causes the weakening of laboratory mixed stabilized soil, which is thought to be one of the reasons for the high strength ratio of the field strength q_{uf} to the laboratory strength q_{ul} (Sasaki et al., 1996; Ishibashi et al., 1997). In order to prevent the separation, mix the stabilized soil by hand in a mixing bowl and scoop it into the molds quickly.

In the case of a uniform sandy soil with less fines content being mixed with a slurry form binder, excessive tapping of the mold for air removal may cause the density and strength to decrease. In the case of loam, or a clayey soil with sand, being mixed with powder form binder, the mixture can form lumps during mixing by an electric mixer. If it happens, the lumps should be broken up before it is placed in molds.

3.3 Curing

The specimen in the mold is covered with a sealant to prevent a change in water content, and cured at $20 \pm 3°C$ over a prescribed curing period. The curing period

(a) Temperature-and humidity-controlled container.

(b) Humidity-controlled box.

Figure A.6 Examples of a curing container and a box.

may be selected from 1, 3, 7, 14, 28, and 91 days, etc., depending on the purpose of the test and the type of binder. It is common and desirable to include 7 and 28 days.

The following are desirable curing procedures: (1) sealed mold and/or specimen should be placed in a temperature- and humidity-controlled container (Figure A.6(a)); (2) sealed mold and/or specimen should be placed in a humidity-controlled box (relative humidity above 95%) and the box should be placed in a temperature controlled room (Figure A.6(b)). The utmost care should be paid to prevent tears in the sealant material, to assure tight sealing. The reason for not curing the specimen directly underwater is that the actual stabilized soil is mostly cured underground with negligible migration of water.

3.4 Specimen removal

Once the strength of the stabilized soil specimen is found to have reached a sufficient level, the specimen may be taken out of the mold to complete the curing process. Figure A.7 shows an example of removal of a specimen by splitting the lightweight plastic mold along pre-processed slits. The exposed end of the specimen must be trimmed properly before removing it from the mold. The removed specimen should be put in a polyethylene bag or wrapped in a sheet of high polymer film (such as food storage-type plastic wrap) and placed back in the curing container to complete the curing process. To avoid possible deformation due to excess load, the wrapped specimens should not be stacked.

4 REPORT

In the report, it is desirable to report both the binder factor as well as the binder content, as they are most commonly used. There are other elements to do with the binder amount that should be reported, such as (1) the ratio of the dry weight of binder to the wet weight of soil, and (2) the ratio of volume of binder slurry to the volume of soil.

Figure A.7 Removal of a specimen of stabilized soil from a lightweight plastic mold by splitting it along pre-processed slits.

Figure A.8 An unconfined compression test on stabilized soil.

The type and amount of chemical additives should be reported, if used. Also, it is desirable to report any data on the amounts of all materials such as soil sample and binder measured during the preparation procedure. Table A.1 shows an example format for the specimen preparation report (Japanese Geotechnical Society, 2009).

Table A.1   Example format for a test report.

Specimen condition	Binder			Admixture*		
	Binder factor (%)			Admixture amount*** (%)		
	Binder content (kg/m³)			Number of specimens		
	Slurry mixing water type*			Mold dimension (volume)		ϕcm × cm (cm³)
	Water/binder ratio* (%)					
	Curing period (days)					
Raw soil water content	Container No.					
	m_a (g)					
	m_b (g)					
	m_c (g)					
	w (%)					
	Average $\overline{w}$ (%)					
Required amount of materials per one batch		Soil, m_T (g)			Binder, m_H (g)	
		Water*, m_W (g)			Admixture*, m_A (g)	
Stabilized Soil water content	Container No.					
	m_a (g)					
	m_b (g)					
	m_c (g)					
	w (%)					
	Average $\overline{w_s}$ (%)					

For saturated soil, determine the required material amounts by the equations below.   Note:

① Soil mass, m_T (g) :

$\qquad\qquad\qquad\qquad\overline{w}\qquad\qquad\qquad\rho_s\quad\overline{w}$

Wet density, ρ_T (g/cm³) = {1 +/□} {1/+/100} □ □

=

m_T (g) =×□   □   □   □

# of specimens volume of mould ρ_t extra**

② Binder mass, m_H (g):

$\qquad\qquad\qquad\qquad m_T\qquad\qquad\overline{w}$

Soil dry mass, m_D (g) = / (□100)= □   □

m_H (g) = □0 = □   □

$\quad\quad m_D$ binder/soil ratio

Binder content (kg/m³) = {10 ρ_T / (1 + $\overline{w}$/ 100)} × Binder factor

③ Mass of slurry mixing water, m_W (g)

m_W (g) = ;□0 = □   □

$\quad\quad m_H$ binder/soil ratio

④ Admixture mass, m_A (g)   Notes:

m_A (g) =×□□ = □   □   *   only if used

$\quad\quad m_H$ amount   **   ratio w.r.t. binder mass

*** normally 1.1~1.2

5   USE OF SPECIMENS

The stabilized soil specimens are mostly used for unconfined compression tests. However, they can also be used for triaxial tests, simple tensile strength tests, splitting tensile strength tests, cyclic triaxial tests, and fatigue strength tests.

Table A.2 Unconfined compressive strength of various stabilized soils.

Soil									Binder			Unconfined Compressive Strength[6] q_u (kN/m²)	
			Grain Size Composition										
Sample Location	Soil Type	Water Content (%)	Sand (%)	Silt (%)	Clay (%)	Liquid Limit (%)	Plastic Limit (%)	Organic Content[1] (%)	Type[2]	Powder/Slurry[3] (W/C[4])	Binder/Soil Ratio[5] (Amount of binder[5])	7 days	28 days
Yokohama Bay	Marine Soil	97.9	6.4	37.5	56.1	95.4	32.3	3.6	NP	C slurry (60 %)	13.5 (100)	2,140	2,870
									BF		13.5 (100)	1,180	1,990
									NP		27.0 (200)	4,050	5,490
									BF		27.0 (200)	3,690	5,960
Osaka Bay	Marine Soil	93.9	3.5	30.8	65.7	79.3	30.2	2.7	NP	C slurry (60 %)	13.1 (100)	950	1,400
									BF		13.1 (100)	980	1,470
									NP		26.2 (200)	1,490	2,750
									BF		26.2 (200)	3,150	4,890
Imari Bay	Marine Soil	83.3	2.2	44.5	53.3	70.4	24.2	4.3	NP	C slurry (60 %)	12.0 (100)	540	830
									BF		12.0 (100)	490	830
									NP		24.0 (200)	1,130	2,060
									BF		24.0 (200)	2,190	4,250
Tokyo Prefecture	Land Soil	54.0	5.0	53.0	42.0	44.7	23.9	3.8	NP	C slurry (80 %)	4.6 (50)	530	730
									BF		4.6 (50)	160	350
									NP		6.8 (75)	1,260	1,760
									BF		6.8 (75)	580	1,090
									NP	CB slurry (200%)	22.8 (250)	700	1,510
									BF		22.8 (250)	1,110	2,410
Funabashi, Chiba	Land Soil	14.2	95.6	3.1	1.3	–	–	–	NP	CB slurry (80%)	15.3 (300)	460	910
									BF		15.3 (300)	560	1,800
									Slag		15.3 (300)	1,110	2,860
Fujishiro, Ibaragi	Land Soil	236	–	–	–	251	92.7	25.2	NP	C slurry (80%)	72.5 (250)	130	190
									BF		72.5 (250)	140	160
									For Organic Soil		72.5 (250)	490	780
Nangoku, Kouchi	Land Soil	295	–	–	–	272	69.1	17.6	NP	C slurry (80%)	85.0 (250)	140	250
									BF		85.0 (250)	98	200
									For Organic Soil		85.0 (250)	590	1,570
Haneda	Reclaimed Land Soil	160	1.0	33.0	66.0	99.1	39.7	4.8	Quicklime	Powder	10 (–)	540	740
									Quicklime		20 (–)		
Yokohama	Reclaimed Land Soil	102.5	9.9	44.6	45.5	78.8	39.1	2.95	Quicklime	Powder	10 (–)	1,670	2,740
									Quicklime		20 (–)	2,350	3,720
Naruo, Hyogo	Marine Soil	90.2	2.0	26.1	71.9	83.0	31.4	–	Quicklime	Powder	10 (–)	250	690

Notes: 1) Organic contents of soil are determined according to JGS T 231 'Testing Procedure for organic content of soil' (chromic acid oxidation method). 2) NP: ordinary Portland cement; BF: blast furnace cement type B. 3) C slurry: cement slurry; CB slurry: cement-bentonite slurry. 4) W/C: water/cement ratio. 5) Binder/Soil ratio (%): ratio of binder mass to dry soil mass; Amount of binder: binder mass (kg) per m³ of test soil. 6) The unconfined compressive strengths of stabilized soil with quicklime is obtained from the figures (Terashi et al., 1997).

REFERENCES

Babasaki, R., Terashi, M., Suzuki, K., Maekawa, J., Kawamura, M. & Fukazawa, E. (1996) Factors influencing the Strength of improved soils. *Proc. of the Symposium on Cement Treated Soils.* pp. 20–41 (in Japanese).

Ishibashi, M., Yamada, K., & Saitoh, S. (1997) Fundamental study on laboratory mixing test for sandy ground improvement by deep mixing method. *Proc. of the 32nd Annual Conference of the Japanese Geotechnical Society.* pp. 2399–2400 (in Japanese).

Japanese Geotechnical Society (2009) *Practice for making and curing stabilized soil specimens without compaction. JGS 0821-2009.* Japanese Geotechnical Society. Vol.1. pp. 426–434 (in Japanese).

Japanese Industrial Standard (2006) *Method of making and curing concrete specimens, JIS A 1132: 2006.* (in Japanese).

Kitazume, M., Nishimura, S., Terashi, M. & Ohishi, K. (2009) International collaborative study Task 1: Investigation into practice of laboratory mix tests as means of QC/QA for deep mixing method. *Proc. of the International Symposium on Deep Mixing and Admixture Stabilization.* pp. 107–126.

Terashi, M., Okumura, T. & Mitsumoto, T. (1977) Fundamental properties of lime-treated soils. *Report of the Port and Harbour Research Institute.* Vol.16. No.1. pp. 3–28 (in Japanese).

Subject index

additive, 6, 9, 24(Table 2.1), 220, 224
  type of, 18
admixture stabilization, 1, 2, 23, 60, 110
  classification of, 3(Table 1.1)
air pressure feed system, 176, 177
  mixed air feed method, 176, 177
  pressurizing pump method, 176, 177
  pressurized tank method, 176, 177
alkali, 25, 116, 138, 201
axial strain at failure, 70, 81, 134, 135,
  208

backfilling, 11, 18, 154, 190
bearing capacity, 4, 205, 208, 213
  ultimate bearing capacity, 133
  yield bearing capacity, 133
bearing capacity factor, 213
bending failure, *see failure mode*
beneficial use of dredegd soil, xi, 1, 3,
  18, 143, 166
binder, xii, 1, 2, 3, 68, 218, 220,
  binder content, 18(Table 1.3), 33, 42,
  220
  binder factor, 14, 18, 42, 143, 220
  binder injection system, 179, 181
  binder injection technique, 175
  binder supplier barge, 174(Table 5.1)
  binder supplier system, 177
  characteristics of binder, 24(Table
  2.1), 33,
  chemical composition of binder, 25,
  32(Table 2.5),
  influence of characteristics of binder,
  25
  minimum binder content, 42

soil and binder mixture, 68, 133, 174,
  187, 220
type of binder, 28, 36, 113
binder slurry, 28, 44, 191, 220, 223
binder supplier barge, 174(Table 5.1)
binder supplier system, 177
  compresssor addition type, 174, 177,
  178
  line addition type, 174, 177, 178
blade, 218,
  mixing blade, xi, 4, 133
buggy, 192, 197

calcium
  calcium content distribution, 97, 98,
  99
  calcium hydroxide, 24, 116
  calcium oxide, 26
  calcium leaching, 98
cement, 40, 41(Table 2.4), *see also
  binder*
  blast furnace slag cement, 18, 24,
  26(Table 2.2), 28, 33, 52(Table 2.7)
  cement-based special binder, 25,
  26(Table 2.3), 28, 109, 113, 147
  ordinary Portland cement, 18, 24,
  26(Table 2.2), 28, 33, 52(Table 2.7)
cement content, *see binder content*
cement factor, *see binder factor*
cement hydration, 2, 24, 47, 59, 70, 75,
  76, 120
cement stabilization, xi
  mechanism of cement stabilization, 2,
  24

chemical additives, 6, 220, 224, *see also additive*
coefficient of variation, 128, 129, 151, 157, 161, 162, 164, 208
column, *see stabilized soil column*
cone penetration resistance, 8, 131, 139, 170, 197
consistency, 35, 186, 219, 220
  change in consistency, 2, 68, 79, 117, 133, 135, 138,
consolidation characteristics, 70, 101, 137
  coefficient of consolidation, 75, 104, 137
  coefficient of permeability, 75, 106, 137
  coefficient of volume compressibility, 75, 103, 137
  consolidation yield pressure, 75, 82, 101, 137
COV, *see coefficient of variation*
creep strength, 87, 136
curing condition, 24(Table 2.1), 46, 92
  influence of curing temperature, 51, 54
  influence of curing period, 47, 52(Table 2.7), 54, *see also strength ratio*
  influence of maturity, 53
cyclic strength, 87, 136

deep mixing method, 4
density, 77, 110, 184, 191
  change of density, 77, 135
design strength, 197, 205, 206, 207
deterioration, 93, 97, 98, 136, *see also long-term strength*
  deterioration depth, 95, 96, 98, 136
  exposure test, 98, 99
  strength decrease, 93, 136
dewartering, 3(Table 1.1)
dioxin, 116,
drying and wetting cycle, 54
dynamic property, 85, 136
  damping ratio, 87
  initial shear modulus, 85, 136

earthquake
  liquefaction prevention, 6, 213
earth pressure, 205, 208, 211
  dynamic active earth pressure, 210
  earth pressure at rest, 209
  static active earth pressure, 209
elution
  elution criteion, 109(Table 3.1)
  elution of contaninet, 108, 137
  arsenic, 109(Table 301), 110(Table 3.3), 137
  boron, 109(Table 301), 110(Table 3.3), 137
  cadmium, 109(Table 301), 110(Table 3.3), 137
  fluorine, 109(Table 301), 110(Table 3.3), 137
  hexavalent chromium, 109(Table 301), 110(Table 3.3), 113, 138, 191, 201
  lead, 109(Table 301), 110(Table 3.3), 137
  mercury, 109(Table 301), 110(Table 3.3), 137
  selenium, 109(Table 301), 110(Table 3.3), 137
  trichloroethylene, 109(Table 301), 110(Table 3.3), 137
exposure conditions, *see deterioration*
ex-situ mixing, 3(Table 1.1), 6

failure mode, 212, 213
  bending failure, 213
  shear failure, 213
  tensile failure, 213
field strength, 60, 62, 68, 205, 222
  ratio of field strength and laboratory strength, q_{uf}/q_{ul}, 128, 151, 154, 156, 161, 164, 169, 208, 222
  variability of, 68, 206, *see also coefficient of variation*
field trial test, 186
flow value,14, 16, 122, 125, 138, 151, 156, 164, 165
  change in, 69, 134
  design flow value, 191
  target flow value, 156, 186

geotechnical design, 68, 173, 205, 206, 208

grain size distribution
 influence of grain size distribution, 35, 108, 219
 sand fraction, 35,

hazardous substance, 109(Table 3.1), 110(Table 3.2, Table 3.3)
heterogeneity, 129, 139
high pressure injection mixing, 3(Table 1.1)
humic acid
 influence of, 35, 36
hydration, *see cement hydration*

ignition loss, 36, 191
improved ground, 173, 205
influence of type of water, 32
initial shear modulus, *see dynamic property*
in-situ mixing, xi, 3(Table 1.1), 4
 deep mixing, *see deep mixing method*
 mid-depth stabilization, 3(Table 1.1), 4
 surface and shallow stabilization, 3(Table 1.1), 18, 67, 143
internal friction angle, 81, 135
ion exchange, 2, 68

laboratory mix test, 28, 43, 44, 62, 191, 197, 206, 217
laboratory strength, 60, 62, 68, 205, 206, 222
land reclamation, xi, 1, 10, 18, 150, 154, 155, 160, 164, 166, 189
lightweight treated soil method, 6
lime, 4, 26, 67, 81, 85
 hydrated lime, 85, 104, see also calcium hydroxide
 quicklime, 68, 84, 122, *see calcium oxide*
lime stabilization, 59, 68, 122
liquefaction
 liquefaction prevention, *see earthquake*
 liquefaction resistance, 3, *see also earthquake*

long-term strength, 26, 92, 94, 95, 98, 137, 157
strength increase, 94, 136
strength decrease, *see deterioration*

marine construction, 11, 193, 220
maturity, *see curing condition*
mechanical mixing, xi, 3(Table 1.1)
mixing conditions, 23, 24(Table 2.1), 42, 69, 147, 191, 206
mixing time, 43, 45
mixing degree, 122, 151, 175, 180,
 degree of mixing, 27(Table 2.1), 123, 152, 178
mixing ratio, 13
modulus of elasticity, 84, 135
molding, 44, 217, 221,
 remolding, 186

narrow multi-beam echo sounder, 192, 197

on-land work, 11, 154, 193
organic matter content, 23, 161, 170, 217, 219, *see also ignition loss*
overburden pressure, 6, 55, 161, 213
overlap, 220

pipeline, xi, 1, 3(Table. 1.1), 10, 147, 156, 174(Table 5.1), 176, 180, 186
 airpressure distribution in, 13, 16, 165
 inner surface of, xii, 11
 pipeline diameter, 13, 17, 165, 174, 194(Table 5.2), 195(Table 5.3)
 pipeline length, 13, 125, 156, 174, 191,
pneumatic flow mixing method, 23
 development of, 10, 173
 statistics of, 20
 drum mixing method, 181
 Kokusou-diagonal pipe slurry mixing (K-DPM) method, 181
 line mixing pneumatic conveying system (LMP) method, 146, 181
 Mitsui in line dispenser (MILD) method, 181
 pipe mixing method, 181, 184, 185

pneumatic flow mixing method,
    (*Continued*)
  plug magic method, 146, 179, 181,
    183,
  snake mixing method, 146, 151, 152,
    180, 182
  W-tube mixing method, 181, 182
permeability, *see consolidation*
    characteristics
pH, 97, 116, 138, 191, 197, 202, 219
  influence of pH, 24(Table 2.1), 38, 39
placement, xii, 24(Table 2.1), 70, 128,
    188, 198, 206, 214
  effect of placement, 127, 139
  on land placement, 192, 197
  under water placement, 127, 156, 184,
    193, 197
  placement barge, 174(Table 5.1), 186,
    188, 190
  placement control, 192
  placement equipment, 173, 180, 184,
  placement machine, 101, 162, 167
placement method
  cyclone placement method, 147, 188,
    189
  direct placement method, 184, 188,
    190, 194(Table 5.2), 195(Table 5.3)
  tremie placement method, 189, 190
placement procedure
  placement barge procedure, 189, 190
  slope shoulder flow-down procedure,
    189
  tremie barge with conbtrolled water
    level procedure, 189
plasticity index,69, 79, 133, 135
plug flow, 12, 13, 183
plug soil, xii, 12, 13, 17(Table 1.2), 125,
    126, 175, 176, 183, 187
Poisson's ratio, 85, 136
power blender method, 4
pozzolanic reaction, 2, 24
pozzolanic reactivity, 28
premixing method, 6

quality control, xii, 10, 125, 151, 155,
    156, 167, 173, 184, 191, 197, 201,
    217

quality assurance, xii, 151, 155, 173,
    197
q_{uck}, *see design strength*
q_{uf}/q_{ul}, *see field strength*
quicklime, *see lime*
q_{u28}/q_{u1}, 52(Table 2.7)
q_{u28}/q_{u3}, 52(Table 2.7)
q_{u28}/q_{u7}, 52(Table 2.7)
q_{u28}/q_{u7}, 49, 52(Table 2.7)
q_{u91}/q_{u7}, 49, 52(Table 2.7)
q_{u91}/q_{u28}, 49, 52(Table 2.7)

remolding, *see molding*
residual strength, 70, 84, 120, 135, 138
resolution
  alkali, 116, 138
  dioxin, 116, 138
rest time, 44, 45

sand fraction, *see grain size distribution*
sand particle content, 18, 150, 186, 195
sea reetment, 11, 18, 143, 156, 190, 198
sea water, *see influence of type of water*
secant modulus of elasticity, 84, *see also*
    modulus of elasticity, 84, 135
secondary compression, 108, 137
seismic coefficient, 211, 212
seismic force, 211
settlement
  settlement reduction, 166, 167
shallow mixing, 3(Table 1.1), 94, 220
shear failure, *see failure mode*
soil compaction, 24(Table 2.1), 56, 59,
    117, 121, 138
soil disturbance, 24(Table 2.1), 56, 59,
    117, 121, 138
soil plug, *see plug soil*
soil violume design, 205, 213
stabilized soil column, 4, 94, 220
stabilizing agent, *see binder*
standard deviation, 45, 207
strain at failure, *see axial strain at failure*
strength ratio
  q_{uf}/q_{ul}, *see field strength*
  q_{u28}/q_{u1}, 52(Table 2.7)
  q_{u28}/q_{u3}, 52(Table 2.7)
  q_{u28}/q_{u7}, 52(Table 2.7)

q_{u28}/q_{u7}, 49, 52(Table 2.7)
q_{u91}/q_{u7}, 49, 52(Table 2.7)
q_{u91}/q_{u28}, 49, 52(Table 2.7)
stress–strain curve, 70, 79, 84, 119
surface stabilization, 3(Table 1.1)
suspended solids, 198, 199, 201

tap water, *see influence of type of water*
tensile crack, 212,
tensile strength, 90, 91, 136, 225
tensile strength test, 91, 225
   split tension test, 91
   simple tension test, 91
   bending test, 91
transportation distance, 13, 17, 125,
   126, 139, 148, 152, 165, 178
   minimum transportation distance,
   180, 191
tremie pipe, 162, 174, 183, 184, 189,
   190
turbidity, 197, 201
turbulent flow, xii, 1, 12, 17, 187

unconfined compressive strength, 18,
   84, 91, 131, 143, 194, 206, 217
unit weight, 6, 9, 78, 124, 184, 187,
   192, 209
   change in unit weight, *see change in
   density*

void ratio – consolidation pressure
   curve, 75, 101, 134, 137

water content, 2, 13, 18(Table 1.3), 125,
   157, 186, 187, 191, 220
   change in water content, 76, 77, 135
   influence, 40, 42, 61
water quality control, 201
water to binder ratio, 179, 191
water to cement ratio, W/C, 41, 43, 61,
   see also water to binder ratio

Young's modulus, *see modulus of
elasticity*

Milton Keynes UK
Ingram Content Group UK Ltd.
UKHW051853071024
449327UK00025B/1942

9 780367 574246